W9-BVB-779

A Sense of Self

A Sense of Self

Memory, the Brain,
and Who We Are

VERONICA O'KEANE

W. W. NORTON & COMPANY
Independent Publishers Since 1923

For Esther and Sean

First American Edition 2021

Originally published in Great Britain under the title *The Rag and Bone Shop: How We Make Memories and Memories Make Us*

Manufacturing by Lake Book Manufacturing

Library of Congress Cataloging-in-Publication Data

Names: O'Keane, Veronica, author.
Title: A sense of self : memory, the brain, and who we are / Veronica O'Keane.
Description: First American edition. | New York : W.W. Norton & Company, 2021. | Includes
    bibliographical references and index.
Identifiers: LCCN 2021005133 | ISBN 9780393541922 (hardcover) | ISBN 9780393541939 (epub)
Subjects: LCSH: Brain. | Memories. | Neurosciences. | Mind and body.
Classification: LCC QP376 .O34 2021 | DDC 612.8—dc23
LC record available at https://lccn.loc.gov/2021005133

W. W. Norton & Company, Inc., 500 Fifth Avenue, New York, N.Y. 10110
www.wwnorton.com

W. W. Norton & Company Ltd., 15 Carlisle Street, London W1D 3BS

1 2 3 4 5 6 7 8 9 0

Contents

Contents

Foreword

This heart within me I can feel, and I judge that it exists. This world I can touch and I likewise judge that it exists. There ends all my knowledge and the rest is construction.

Albert Camus, *The Myth of Sysiphus* (1955)

A small detail about the translation of the title of Proust's famous exploration of his childhood memory, *À la recherche du temps perdu*, demonstrates much of what I will address in this book. Initially translated in 1954 as *Remembrance of Things Past*, the English title was changed in the 1992 edition to the more accurate *In Search of Lost Time*. The original translation's 'remembrance of' suggests a passive recall of memories from a hidden and fixed repository, while the later translation's 'in search of' suggests an active pursuit of a fluid past that is lost. Neuroscience almost caught up with Proust in the interval between the translations.

Notes

Literary bibliographical references and short explanatory notes are given in footnotes. Discursive notes are indicated by roman numerals, and can be found on pp. 235–47. Academic and scientific references are indicated by arabic numerals, and can be found on pp. 249–60.

PART ONE

How We Make Memories

Memory is the medium through which we filter present experience and create a sense of time, place and person. Through my own experiences as a clinician and scientist and through the experiences of my patients, we will see how neural pathways carry sensations from the outside world to become representations in the brain, and the role that emotions play as an intrinsic part of the memory process. We then dive into the brain's memory factory of the hippocampus, the emotional fire-spark of the amygdala and the rag and bone shop of the insula to understand the inner workings of the brain. Finally, I will explore why we fail to even register some things, while vividly remembering others.

1.   Dawnings

There are events that occur in each of our lives that are experienced with a prescient sense that they will always be remembered. Occasionally this feeling is particularly intense and, while not epiphanic, carries with it a sense of having stumbled into a new level of awareness. This new awareness is pre-verbal, like the rattling of cups on saucers that is the only sign of subterranean shifts. The rattle that set me on the journey to understanding the real matter of memory occurred in London in the early 2000s. Looking back, the incident is like the introductory scene in a novel where every ingredient of the story to be told is laid out with a canny casual innocence that, if retrospectively analysed, is prophetic of the outcome. Edith's story set me on a journey of breaking down and reformulating my ideas about memory – knowledge that had become automatic to me, but that somehow eluded the material essence of what it is to be a sensate, living human, with a memory sculpted by individual experience.

I met Edith in the Bethlem Royal Hospital, the oldest psychiatric hospital in the world and now part of the Maudsley Hospital of more contemporary fame. The Bethlem dates back to 1247, when it was called the Bedlam, until 'bedlam' became a noun that indicated chaos and turmoil. The hospital was renamed in the early twentieth century as the Bethlem Royal. National treatment units were located in the 100-plus acres of horse chestnut and hazel tree grounds of the hospital. I worked for five years as the lead clinician in a National Perinatal Psychiatric Unit in the early 2000s, a unit that has so far been spared the decimation in NHS services that has since occurred. Women from all over the UK were transferred to us for specialist treatment of perinatal psychiatric illnesses – illnesses that arise during pregnancy or during the postpartum period.

A badger family had setted in a tunnel in the ground close to the entrance to our unit. I often paused to look at the opening to their sett in its soft grassy mound on the off chance that the badger would, perhaps in a spontaneous moment of protective night watchmanship, pop its head out during daylight. I was commuting between London and Dublin during these years, and my two little ones in Dublin waited each week for news of sightings, but they had to do instead with pressed woodland flowers in spring and summer, and hazelnuts and conkers in late autumn. I loved my five years of working in the Bethlem, bringing women like Edith, who had been felled by the cruel illness of postpartum psychosis, back to their lives. Most of the women admitted to our unit had this rarely spoken-about psychosis, which occurs in about 1,400 women in the UK yearly. Edith was admitted to the Bethlem a few weeks following the birth of her baby. This is her story.

Edith had no history of psychiatric illness when she gave birth to her baby. Baby's arrival was anticipated with joy. The pregnancy was healthy, and fetal scans were normal. Labour was uncomplicated and a healthy baby was delivered at term. In the days following Baby's birth, Edith became emotionally distant, and seemed to become increasingly bewildered. She appeared to be distressed and preoccupied but would not communicate the cause of her distress. Her condition rapidly deteriorated and at the time of her admission she had stopped eating and was wandering aimlessly around her home day and night ignoring Baby and the rest of the world. Her family doctor assessed her at home and she was referred immediately to us for assessment and treatment. When I met Edith I noted that she was unusually slim in spite of having given birth less than two weeks previously. She had a closed expression and was more or less mute and unresponsive to our questions.

We frequently see this 'locked-in' mien in individuals with psychotic experiences. In the case of women with postpartum psychosis, they usually hear voices that are not audible to others, may smell odours – usually unpleasant – that are not coming from the outside world and may feel sensations on their body that are not caused by anyone or anything that is visibly touching them. Such auditory, olfactory (smell), visual or somatic (touch or visceral) hallucinations are referred to as psychotic symptoms. The first principle that we need to establish is that what are called symptoms are real sensory experiences. Hearing a sound, a human voice, is a subjective experience, whether the voice originates in the outside world or is generated in the brain by pathological neuronal firing. The experience of hearing the voice is similar in both cases: the origin of the sensation is a separate consideration. If the experience is generated by pathological brain-firing, the voice hearer will look around to see who's talking, as one does, and may attribute the voices to whoever is present, or to hidden speakers. Commonly, those experiencing auditory hallucinations will look as if they are talking to themselves, whereas in their reality they are responding to voices that are as audible and real to them as the voice of a living person.

This leaves the psychotic person isolated, trapped in a sensate world that is a misinterpretation of the outside world. They may come to believe that they are privy to a level of sensory experience that is not available to others, a 'sixth sense'. Mostly, individuals in a psychotic state will invoke unseen forces, such as IT, ghosts, magic, deities or, in the case of Edith, the devil, to explain these subjective experiences that are at odds with the world as experienced by others around them.

Edith was preoccupied with making sense of her vivid experiences and was unable to respond to the world of external sensate stimuli. Like most women in the throes of postpartum psychosis she appeared to be in a state of altered consciousness, as if removed from the world. On assessment I noted that Edith sometimes stared into my eyes and sometimes firmly closed her eyes shut, and occasionally she stared at a member of the team. She seemed to be looking at whoever was placed in the direction of the voices that she was hearing. Her movements were stilted and

5

non-purposeful. She was guarded and trying to hide her confusion and her fear. It was apparent to us that Edith was responding to sensory stimuli that were not originating from the external world, that she had a postpartum psychosis.

Edith had stopped caring for Baby. She 'knew' that her baby was not the baby that she had given birth to, although it looked identical. Her own baby could not have a rotting smell. So Baby had been switched, somehow. In the beginning she was not sure whether her birth baby had been taken away and the baby before her was an identical substitute, or whether her baby had been taken over by some evil spiritual force, probably the devil. On the way to the Bethlem she passed a graveyard that was familiar to her, being close to her home. Glancing through the gates, her eye caught a small gravestone that she noticed had a slight tilt. She suddenly realized when she saw the small gravestone that her baby had been buried there. The old headstone disguised the new burial, and it was tilted because it had been recently disturbed. This proved that what she now had was an imposter baby. The evil separation of Edith from her birth baby was complete, and now she was being locked up by the perpetrators of the malevolence.

She did not reveal this to me or to anyone else when she was admitted to the hospital, because that would have given the game away, thereby exposing herself. It was only by pretending that she did not know that we were feigning roles to deceive her, that she would have a chance to save herself. She could give nothing away. She was playing our game and trying to say as little as possible.

One of the experiences that I have frequently observed in women suffering with postpartum psychosis is a belief that people close to them, and particularly their newly delivered babies, have been replaced by a double, an imposter. This phenomenon is called Capgras syndrome, after the physician who apparently first described it. I say 'apparently' because the idea of changeling babies goes back as far as our oldest stories, the fairy tales. We'll return to the fairy story at the end of the book.

*

Apart from Baby, Edith also thought that her partner was an imposter, an identical substitute, colluding in the plot to harm her. She only revealed this to me months later, following her recovery. Because Edith was terrified at being taken captive by evil forces, she wanted to escape from the hospital. She refused to take medication, which she predicted would be poisonous, or at best a drug that would weaken her power to fight the conspiracy. She reasoned that she was the only one left to dispose of before the new order could be established. Her imposter husband and the grotesque production around her were now targeting her. Gestures among the malicious schemers carried meaning; nothing was accidental or incidental. No one was who they appeared to be, and her imposter family, in collusion with others, had taken her baby away, killed him and buried him in haste in the local graveyard.

We decided that it would be unsafe for Edith to leave the Unit, and treatment with antipsychotic medication was commenced. Over the course of days she became less distressed and began to respond to us. Two weeks later, as the psychosis receded, she became distressed at being separated from what she now realized to be her birth baby and wanted to be reunited with him. When Baby was brought to the Unit by Edith's partner she responded with tears and joy. I can't imagine the confusion of emotions that she was experiencing, but among them were the emotions of a newly delivered mother. She gradually recovered and left our Unit three weeks later, no longer psychotic but traumatized by what had happened to her.

On return visits to my outpatient clinic over the subsequent months, Edith told me about what she had been experiencing when psychotic. Following the commencement of treatment the voices had gradually faded from a normal volume to whispers, then became less frequent and had finally faded away. The thoughts that her partner and baby had been replaced faded also, and with this the idea that everyone around her, including the medical team, were part of a paranoid plot. She became quite shamed by her beliefs when psychotic, particularly about Baby, and wanted to put the whole episode behind her. She was also worried that, if she revealed what she had thought had been happening, others might see her as an unsafe mother. Before becoming psychotic, Edith knew little about psychosis and had never heard the term postpartum psychosis. Her understanding of herself had been turned upside down. I reassured her that the psychosis was an illness caused by the rapidly changing hormones during birth that had affected her brain . . . that this had caused parts of her brain to fire off, creating subjective experiences that seemed to come from the outside but that had really been generated within her brain.

Subjective experience is where any explanation of psychosis must begin. All sensations, whether a voice, a smell, a touch, a visual image, whether 'psychotic' or 'real', whether being stimulated by something in the outside world or because the brain is firing off for no apparent reason and without external sensation, are experienced as real. Edith and I had established that her experiences had been subjectively perceived as real experiences, and were therefore unmistakably subjectively real. We would refer to the experiences as real, with an implicit understanding that they were also psychotic.

The scene that came back to me again and again was a conversation that we had following her discharge. I asked her if she had experienced fleeting psychotic ideas about Baby or her partner since her discharge. Edith responded that she had done so in the early stages of recovery, but less so as time progressed. She told me that as she was passing the graveyard on her way to an outpatient consultation following discharge from the unit, her eye caught the small gravestone in her local graveyard that she had seen on her way to involuntary detention in the Bethlem. This was the same gravestone that she had immediately understood then, prior to admission, as being where her baby was buried. As she looked at the small tilted gravestone months later, for a few moments she was 'back' on her way to the hospital being involuntarily detained by the imposters who had replaced the real people in her life. She had a rush of the full range of these beliefs, accompanied by a feeling of terror. I asked her if she knew that the psychotic ideas were not real on this, the second, occasion. What she said next set me on a long-term pathway of inquiry about the nature of the matter of memory. She looked straight at me and said, 'Yes . . . but the memories are real.'

And so I learned that Edith's memory seemed to exist as a

discrete organic entity – an experiential snapshot, a 'flashback'. What *is* a flashback except a vividly experienced memory? For Edith, the time interval between the event and the recall had disappeared, and the memory was a present lived experience hitting her with an emotional punch all over again. The experience of this memory was a thing apart and more powerful than all the reasoning and insight into psychosis that she had accumulated since the memory was laid down. Edith knew that she had been psychotic, knew that her psychosis had been treated and that she was now better, knew that her baby was at home, was not a changeling, was not dead and buried in the local graveyard, and so on, but all this knowledge was suspended while she was experiencing the memory. *The memory was real.*

Edith's Proust-like ability to communicate her memory as unreconstructed sensory experience – visual and emotional and seemingly time-independent – initiated a process in me of unlearning learned constructs. Before this conversation I had not really thought about memory beyond the anatomical circuits learned in medical school, the psychological theories learned in postgraduate clinical training, the mnemonic difficulties that occur in brain illness and that we measure in clinical work, and the neuroimaging and molecular research in psychiatry. Memory was more of an abstract construction, drawn from different knowledge repositories. If Edith had told me that seeing the gravestone had reminded her of her coming to the hospital in a psychotic state and that she had experienced a flashback on seeing it again, I would probably have continued to trundle on with this flat understanding of memory.

So, one of the first, of many, lessons that I learned from Edith was that the theoretical classifications of psychology and the clinical classifications of psychiatry were blinding me to subjective experience. Samuel Beckett, the brilliant observer of anguished human states, beloved of intellectuals, wrote, 'I am not an intellectual. All I am is feeling.' This resonates with me, and in this book I have turned my back on intellectual explanations and eschewed theory, even the basic classifications of memory, to follow the journey of memory from sensory experiences of the world and inner feeling states to neural memory lattices.[i]

I have laid out some of the questions, and some possible explanations based on observations of lived experience and scientific experiment, that arose silently over the years that followed, the post-Edith years. How does a visual image trigger a lived memory? How do we re-experience and feel through a memory? What is the difference between a memory that is experienced with emotion and one that is not felt but 'thought', as it were? Why did Edith ascribe the idea of a substitute baby to her strange sensory experiences of hearing voices and rotting smells? If Edith's memory experience of the gravestone as being the burial place of her birth child was a true memory, what then constituted a false memory?

The search through the memory pathways in the brain will show how emotional and feeling states are intrinsically wired in to the laying down of memory and to the experience of recall. We will journey through some of my own professional and biographical memories and hopefully stimulate a slow meandering through some of your own. For thirty-seven years I have observed, treated and researched mood and psychotic disorders. Psychiatrists have a mixed bag of skills – pharmacology, neurology, psychology and intuitions gained by experience – but I think that the expertise that we own exclusively in psychiatry is in the understanding of the nature of experience, what we call 'phenomenology'. We classify some experiences as normal, others as abnormal, and some as pathological. I am not that interested in the distinction between normal and abnormal experience, but I have always had a huge curiosity about the neural mechanisms that *create* experience. One can start anywhere in the search for neural explanations of experience – sensation, cognition or emotion – but it will all eventually lead to memory. Memory brings what we know and what we feel together and becomes the medium through which we filter present conscious and non-conscious experience.

Another fundamental lesson that Edith taught me is that we can learn more easily about normal experience from individuals who have abnormal experiences. William James, a late nineteenth-century psychologist and brother of the more famous novelist Henry, said, 'to study the abnormal is the best way to understand the normal'. So the starting point for me is patients, like Edith, who demonstrate the complexity and entanglements of memory as experienced in real life. I remember patients for many reasons, some for their astonishing resilience and acceptance, others because their presentations have been dramatic or atypical, and others because I could not work out what was wrong. The unexplained presentations remain hovering in my memory, sometimes for many years, until some new understanding dawns, and they suddenly reappear and the puzzle that they posed has been answered. It is as if it is they who have led me to explore, to find and identify the brain mechanism of, their experience. To quote Henry James, brother of the less famous William, 'Our doubt is our passion.'

Edith's gravestone memory, although hidden, was perfectly preserved . . . like the never-seen badger. The badger sett now brings with it, for me, an image of my small children and a sense of the lost opportunities of those precious years that will never return: when time rushed for me, while for them, as for all children, it must have stood still. Personal memory can range from a blindingly sensory and emotional experience, as it was for Edith, to one where there may only be a sense of an emotion – a twinge of sadness, a tiny rush of love, an almost imperceptible knot of loss, a whiff of regret – as I now experience writing this. What does the neural circuitry of memory, which I thought I understood, mean in the world of human experience? This is essentially what I want to explore with you in this book.

2. Sensation: The Raw Ingredient of Memory

In truth, all sensation is already memory.

Henri Bergson, *Matter and Memory*, 1896

The famous short story 'The Yellow Wallpaper' was written by Charlotte Perkins Gilman and published in 1892. Perkins Gilman was a feminist, and the tone of the story is of suppressed gothic horror, reflecting the experiences of life as a woman in the nineteenth century. It can also be read as a fascinating first-person account of a postpartum psychosis. Her genteel descriptions would lead the reader to think that she was the cherished wife of a perfectly kind husband, John, but as the story progresses you realize that she seems to be locked in a disused attic nursery in a largely uninhabited colonial mansion. She doesn't say where this is but tells the reader that she has moved there for the summer, and that she is alone in the nursery. It seemed likely to me when I first read the story that she might be in a psychiatric hospital, because all the windows are barred, the gateway to the stairs is locked, there are chains for restraints attached to the walls, and her bed is nailed to the floor. She is in a state of extreme 'nervousness . . . no one would believe what an effort it is to do what little I am able, – to dress and to entertain, to order things . . . I cry at nothing, and I cry most of the time'. But she meets no one except her husband, who 'is a physician of high standing', and her brother, also a prominent physician, and John's sister, the housekeeper who also tends to her, and whom she refers to as 'Sister', and seems to me to be a nurse. Her baby, 'who she *cannot* be with', is being cared for by Mary. Is she not allowed to be with Baby? . . . is she not able to care for Baby? . . . is she unable to attach emotionally to Baby? . . . is Baby safe with her?

She is allowed to do nothing, having been prescribed total rest, but cheats to write entries in the journal that she is now sharing with the reader. Neither John nor Sister knows about the journal. She becomes obsessed with the patterns in the tattered, intricately patterned yellow wallpaper. 'There are things in that paper that nobody knows but me, or ever will. Behind that outside pattern the dim shapes get clearer every day.' She can see something moving under the wallpaper and can also feel it move, and works out that there is a woman creeping behind it. The creeping woman escapes at night and crawls around the floor. There are also descriptions of her seeing the woman moving in the garden during the day. The wallpaper gives off 'the most enduring odour I ever met . . . the only thing that I can think of that it's like is the colour of the paper. A yellow smell!' For those of you who have not read it, this 6,000-word story can be easily found on the internet.

Like all great works of literature, 'The Yellow Wallpaper' can be interpreted on several levels, all of which have validity. It is a feminist story about women being pasted behind wallpaper, not being allowed to write, being locked up and deprived of sensory stimulation when mentally ill, being treated as hysterical and being treated as a generic group innately intellectually and morally weaker than men, and about the suffocating, patronizing patriarchy of nineteenth-century society and the medical profession. This has made 'The Yellow Wallpaper' a story much investigated by feminist scholars. But . . . Charlotte Perkins Gilman was psychiatrically ill following the birth of her baby, and in a letter to the famous physician Silas Weir Mitchell she speaks about 'the agony of mind [that] set in with the child's coming', her 'dreadful ideas', her 'times of excitement', her days of sleeplessness, of being 'wild, hysterical, almost imbecile at times', and closes with her fear about losing her 'memory entirely', and a request for treatment.[i]

Everything about postpartum psychosis is in this short story: an absent, recently delivered baby; misidentification of the doctors for her husband and her brother; misidentification of the nurse, Sister, for her sister-in-law; the singular and repulsive olfactory hallucination that is ever-present; the visual and tactile hallucinations; the confusion; the sense of her trying to dupe others who are trying to dupe her; the casual references to wanting to burn the house down to get rid of the smell, to biting off a piece of the bedpost, to hiding a rope to restrain the creeping woman when next she escapes . . . The denouement, in which we recognize the creeping woman as the author herself, reflects the disintegration of the sense of self that occurs in psychotic states. 'The Yellow Wallpaper' tells what seems to be a mysterious story in a superficially integrated way and is a brilliant account of a woman trying to bring some coherence, however superficial, to her chaotic sensory hallucinatory experiences.

In this story the woman describes her sensations as they are. She does not sound 'mad'. Her experiences sound strange, but sometimes the world is a strange place. What of her sensations? She *senses* the presence of the woman, she *feels* her behind the wallpaper, she sees her shape in the slithering moving patterns in the wallpaper and finally *sees* her in the flesh after she creeps out from behind the wallpaper, she *hears* her moan and *smells* the horrible 'enduring smell'. These hallucinatory sensations are experienced as real, and we read her diary account of them as being real. Although 'The Yellow Wallpaper' has been interpreted as an account of post-partum psychosis, there has been almost no analysis of the narrator's actual sensory experiences. Her psychotic experiences are generally analysed as a metaphor for her captivity by the harsh societal institutions of the time. It is interesting that, although her experiences are the primary fascination in the story and are what captures the reader, the extensive analysis – you get more than 1 billion hits on a Google search – has almost entirely focused on speculated socio-political meanings of the experiences, rather than the nature of her subjective experiences. She makes sense of her hallucinatory experiences in the same way as we all do . . . we know something because we have seen it, heard it, felt it, smelled or tasted it. The reader knows that there is no creeping woman behind the wallpaper, and yet the narrator does not seem 'mad' in any conventional way. The story possibly shows how close anyone can be to psychosis if placed in a locked room and denied any recognition. The caveat, mostly overlooked by scholars, is that she was probably psychotic prior to being given the 'rest cure' horror of isolation. In the rest of this chapter we will look at how we interpret the world through our senses, and how sensation is the thread that feeds the loom of understanding and memory.

The fundamental point that we cannot make memories without sensation may be so familiar to us that we are blind to it. It is difficult to believe that it took many hundreds of years to understand the now self-evident fact that the five senses bring information to the brain so that one can learn and categorize information and ultimately form a coherent sense of the world. The story of the relationship of sensation to memory dates back to the beginnings of the scientific revolution, four to five centuries ago. The shift in the understanding of memory from being a static repository of knowledge to being a dynamic living human experience is a profound one, and was highly contested. This shift started in the sixteenth and seventeenth centuries, at the time of the beginnings of modern scientific thinking. Copernicus, and then Galileo, proposed that the Earth, rather than being the centre of the universe, was a small planet that travelled around the Sun. This effectively removed Earth from the creationist dogma of the Church.[ii] At that time, the belief systems of the Church had dominated thinking for one and a half thousand years.

The same dogma that denied the science of physics also obstructed the progression of ideas about learning and memory. Men – women were not considered – did not learn about the world through information coming from the world, because it was held to be a truth that all knowledge was given by God and was housed in the soul. There was a God-given soul and a material human body. The idea of a soul, as distinct from a body, has existed for as long as philosophical ideas have been pondered.

Plato's division of humans into body, mind and soul was laid out in the fourth century BC and became an enduring template for the categorization of human experience. The Platonic triad of mind, body and soul was transmuted into comparable Christian triads – for example, God the father, God the son and God the holy spirit/ghost – and has permeated the zeitgeists that have come and gone . . . it is the enduring zeitgeist in successive guises that has had to be endured.

The New Brain/Body Divide: Mind/Brain

The brain/body divides tend to dissolve, and the soul disappears, when the brain causes of mental experience are discovered. A good example of this is General Paralysis of the Insane (GPI), which accounted for up to 25 per cent of admission to psychiatric institutions in the nineteenth century. The symptoms were considered to be a particular type of insanity known as 'moral insanity'. In the 1880s it was discovered that GPI was a brain disease caused by end-stage syphilis, and, following the discovery of penicillin in the 1950s, care of those who suffered from the disease was transferred from alienists, as psychiatrists were then called, to physicians. The identification of the spirochete bacterium that causes syphilis moved the disease from a moral insanity caused by sexual promiscuity to infective medicine. Strange cultural ideas have historically been used to explain mental illness, and this mix of myth and science still continues to confound psychiatry. Epilepsy is an example of a disease that was first treated by psychiatry and was then transferred to neurology – once a cause and treatment for the disorder were discovered. Unexplained mental experience seems to find asylum in psychiatry before being transferred to 'organic' medicine following scientific discoveries.

This is now happening even with psychosis, which is gradually being transferred from a 'mind' to a 'brain' disorder. The mind, in any understanding of the ambiguous word, is the essence of the human brain. The mind is seen as highly subjective and mysterious, but, as we will see, each brain is highly individualized and is forged by the unique experiences and wiring of a person's life. A neat contemporary example of the transfer of a disorder from psychiatry to neurology is NMDA encephalitis.[1] NMDA encephalitis commonly presents with psychotic experiences – voice hearing or paranoia – and movement disorders, and individuals are often admitted and assessed on psychiatric wards. Encephalitis means an inflammation of the brain, and in the case of NMDA encephalitis the inflammation is caused by antibodies directed against brain tissue, often the NMDA receptor found in abundance in the brain. Antibodies are defence proteins produced by the immune system. Usually, antibody synthesis is triggered by foreign organisms, such as a virus, a bacterium or a donor organ, but sometimes the immune system can form antibodies that will attack one's own body tissues, called autoantibodies, and this leads to autoimmune disorders. In the case of NMDA encephalitis, the autoantibody produced by the body fits or matches to the NMDA receptor on neurons and, since NMDA receptors are ubiquitous, brain inflammation – encephalitis – follows. When antibodies form against one's own body – 'autoimmunity' – the targeted tissue becomes damaged because it is identified by the immune system as being a foreign invading body, like a bacterium or virus.[iii] Once the cause of this psychosis was discovered in 2007, it was transferred to neurology. There has been much written since the discovery of this form of psychosis about it being not a psychiatric illness but a neurological one.[2] Since then, evidence has emerged that there is an autoimmune component in many forms of schizophrenia.[3] It is becoming more apparent that brain and mind are one indivisible whole. The transfer of NMDA encephalitis from psychiatry to neurology, even though patients may be presenting with similar experiences and clinical signs to those with other forms of psychosis, is generally viewed as positive by the sufferers. For most, a neurological diagnosis is better than a psychiatric one.

The Third-eye Phenomenon

The roots of the mind/brain divide go back as far as the beginning of recorded history. There has always been some symbolic representation of 'mind' or 'soul', what I call the 'third-eye phenomenon'.

We are at a period in history where we are in awe of the revelations of neuroscience rather than of the extraordinary cultural mythologies that have heretofore explained the experience of being human. My daughter Rowan had a dream one night when she was about thirteen. She woke up in a distressed state, calling for me and for her brother. We sat on her bed as she described vivid staccato scenes from her dream, the sort that one can describe in detail from replaying the visuals immediately on wakening. In the most memorable scene she was in a boat in a large mass of water with a woman, possibly me but she wasn't sure exactly. The boat was rocking in rough waters. Both she and the woman were trying to row towards land that was in sight and looked close, but they could not close the gap because the boat was riding the swells and not making headway. A large snail came out of the sea, terrifying her, and was suddenly in the centre of her forehead. Slowly the snail began to cork-screw into her head following the spiral pattern of its shell, terrifying her into wakefulness.

I was fascinated by Rowan's dream and came to the conclusion that she had transformed the mystical imagery of the 'third eye' into the novel image of a snail. In Egyptian mythology 3,000 years BC this symbol was known as the Eye of Horus. The symbol travelled through the ages, becoming the Eye of Siva in eastern mythology, and in modern day mystic-speak is generically referred to as the third eye. The image is now imbued with vague ideas of unspoken trans-generational female wisdom, having transmuted over the millennia from a masculine symbol of seer-like qualities and protection. Viewed in this context, Rowan's snail could be interpreted as a metaphor of the fears of a female child transitioning – classically a journey over water – into womanhood, being steered precariously over these troubled waters when attacked by the third eye. My interpretation of the iconography may be far-fetched, but I was quite pleased, I have to admit, with her perspicacious fears about her traditional inheritances from female lore and how this could be a destructive invasive force in her thinking.

Rowan's dream did not come from a mystical dipping into an eternal pool of hidden wisdom but seemed to spring from a fear of this mythology. One lesson that I learned from the third-eye dream is that we are steeped in mythology about how we acquire knowledge, and that the imagery is ubiquitous whether you are Christian, Hindu, Buddhist, Muslim or an atheist teenage girl in Ireland. The third eye is now frequently represented as a pine cone, derived from the location of this putative wisdom structure in the pineal gland. The gland was so named because of its similarity to a *pinea* – latin for pine. It is located on a horizontal plane between the eyes, but pushed back behind the brain lobes, similar to the route that Rowan's snail was taking.

In the first century AD the famous physician Galen, possibly the first scientist physician, identified the pineal gland as the probable seat of the soul/mind. He believed, with incredible prescience, that all human experience could be explained by the workings of the body. He didn't believe in an immaterial soul but instead looked to the human brain to explain experience. Galen's suggestion may now sound ridiculous and naïve, but it represented an advance from the common wisdom of the day that the soul and mind were separate – the soul belonging to God, and the mind to the individual.

In the fifteenth century Leonardo da Vinci brought the culturally accepted idea of a soul and a mind together and located the junction in the brain. This nudged the idea of the soul away from one of a wandering spirit with a tenancy in a human body that terminated at the time of death, when the spirit left the body – either to go to an unearthly world or possibly to take up residency in another body – to something that was connected directly to the mind-in-brain. The soul became less spirit and more flesh, less 'god' and more 'man'.

We now know that the pineal gland is quite a primitive brain structure, mostly involved in the secretion of melatonin. Melatonin is important in the world of birds, sheep, horses and cows, in which it is secreted in tandem with available light. It promotes the secretion of their reproductive hormones and ensures the birth of fledglings, lambs, foals and calves during the warm and bright conditions of maximum fertility of land and sea, affording offspring a better chance of survival. In humans, melatonin is involved in the sleep–wake cycle but has no effect of note otherwise. The pineal gland is an unpaired structure in the middle of the brain, which is very rare because almost everything else is paired, and it is tucked into the crevices of deep brain structures like a curious stand-alone vestige of nature-controlled fertility.

The Beginnings of Brain Science

The scientific revolution that exploded post-Leonardo pulled understanding of the world away from the creationist views of the Church and into the universal laws of physics that could explain multiple phenomena. The Earth moved and operated as a consequence, not of the will of God or gods, but of basic laws of physics. The Earth was no longer the centre of the universe, and the idea arose that perhaps man was also subject to some set of scientific principles. The supernatural was being replaced by the natural. In this way science was inadvertently undermining Church dogma.

René Descartes provided a potential solution to the Church/ science divide when he outlined his philosophical principle of *dualism* in the seventeenth century. He argued that the soul was non-matter and God-given and was made of a substance different from the matter of the body – the soul was ethereal and the body was flesh and blood. The brain was a centre of some sort for the material body, but was separate from the immaterial soul. With this theory of 'dualism', Descartes was presenting the first pseudo-scientific explanation of what has evolved to become the

contemporary mind/brain division. He brought together a confusion of ideas about physics and knowledge, a mishmash of the
religious and scientific ideas of the day . . . but importantly he dis-
missed the material sensate experience of living as being of lesser
importance. The invisible soul, perfect and made in God's image
and likeness, was given superiority over the sensate, capricious
body. A group of philosophers vociferously opposed to Descartes'
theory believed that knowledge was accumulated through living,
and specifically through the senses. Those who believed in knowledge being acquired through sensation were called the
Sensationalists. And so we arrive at the origins of the intellectual
battle of whether knowledge (small k) was learned through the
material senses or was innate Knowledge (big K), bestowed in non-
matter by God. Those who opposed Descartes' views, the
Sensationalists, were seen as heretics, and some lost their lives, and
many their liberty, in reclaiming knowledge from god to humans.

We now know that we understand the world through learning,
but in the debate about whether Knowledge was implanted by God
or knowledge was derived from sensate experience, or learning,
there was a lot at stake in terms of the world order. Perhaps the
most important issues were political, because the idea that God
implanted superior Knowledge into certain people, like the pope,
who was infallible, or monarchs, who were chosen by God, gave
absolute power to the Church over lay people, monarchy over the
commoners, men over women, and so on. The various stake-
holders in maintaining this myth of innate superiority fought
within their systems of power. The competing claims of would-be
possessors of innate superiority led to schisms in the Church, the
overthrowing of one exploitative monarch for another, and
Church–state battles. Thousands were slaughtered in inquisitions
and wars. Holding on to the God-given Knowledge idea meant that
knowledge could not be learned to make one person potentially
equal to another.

The battle that was fought throughout the sixteenth and seventeenth centuries remains today, Vesuvius-like, active and simmering,

and sometimes erupting ingloriously. The Sensationalists, although they argued largely from a philosophical and humanist perspective, formed, I think, the intellectual foundations for what are now the disciplines of neuroscience. They were also fighting for human potential over innate Knowledge, and were also the original humanists, laying the foundations for ideals of personal freedom and liberation from tyranny. The story of how *sensation* came to be understood as the material for knowledge and memory is the first chapter in the story of neuroscience, and of modern ideals of human rights.

Molyneux's Question

One of the great debates during the seventeenth century about whether knowledge was innate or was learned through the world of experience is worth elaborating on here, because it shows how arguing about, say, how many angels may dance on the head of a pin can be crisply put to bed with an advance in medical science. The other reason for telling this story is that the debate took place on the campus of Trinity College, Dublin, where I work. The debate was started by William Molyneux (1656–98), a TCD scholar, and John Locke (1632–1704), the famous radical English physician and philosopher. Locke is one of the best-known philosophers of the seventeenth century who opposed the notion of a Cartesian soul and innate Knowledge. He dared to challenge the dictums of the establishment that Knowledge was given by God to men and, furthermore, given to them in proportion to their status. He wrote that the mind was a *tabula rasa* (or blank slate) at birth.* Knowledge was acquired by 'our ordinary abilities to come to know things', as one memorized the world through sensory information. Philosophy encompassed many

*The concept of *tabula rasa* was first proposed by Aristotle in pre-Christian times.

disciplines then, including politics, medicine, psychology, natural sciences, physics and mathematics. Some of the discussions that were held by the Philosophical Society, founded by Molyneux, would be more suited thematically today to the Trinity College Institute for Neurosciences, where I work.

Molyneux wrote a letter to Locke on 7 July 1688 in which he posed a question that has come to be known as Molyneux's Question (or Molyneux's Problem). The question concerned a hypothetical man who had been born blind and had learned to 'see' objects through touch. He was then given sight later in life. The question was whether this man, born without sight and who had learned to distinguish shapes through touch, would be able to distinguish the shapes through looking at them if his vision was restored. He gave as an example a sphere and a cube that the non-sighted man had learned to recognize and differentiate by touch. Would the newly sighted man be able to identify the cube and sphere from sight alone, without touching them?

They approached this puzzle from a philosophical perspective. If the man whose sight had been restored was able to understand the difference between a sphere and a cube through sight alone, without learning shapes through seeing them, then visual knowledge would be already present in his mind, and visual knowledge would be innate. If, on the other hand, the newly sighted man did not know the difference through looking at the objects, this would mean that visual memory was acquired through visual sensory experience, and knowledge was not innate. In this latter empirical framework – that is, learning through observing – one only knew what one had learned through the senses.

Locke and Molyneux correctly reasoned that the blind man would not be able to distinguish between the sphere and the cube *by sight* alone because knowledge was not innate, and had to be learned through each sense, sight and touch, separately. Molyneux's Question was only resolved finally when surgical correction for congenital cataracts, the most common form of congenital blindness, became widely available in the following century. It gradually became apparent that newly sighted people were not able to tell the difference between a cube and a sphere from sight alone, and therefore they did not automatically understand the visual world. The visual image of a sphere and cube had to be learned from touch because that is how the non-sighted patient had learned to *make sense* of objects. Oliver Sacks, the neurologist storyteller, wrote a now famous article in the *New Yorker* in 1993 called 'To See and Not See'. It concerned a man, whom he called Virgil, who was 55 years old when he first gained sight. Everything that Virgil saw for the first time, from his house and its contents to the world of nature, was unintelligible. Sacks noted specifically, in reference to Molyneux's Question, that Virgil could not tell the difference between a sphere and a cube through sight. The mind is a blank slate at birth, and sensate experience of the world accumulates to form knowledge and memory.

Common Sense

The story of Molyneux's Question shows how medical science could translate unanswerable philosophical debates into unequivocal real-life knowledge. This was, as I see it, the first major victory of neuroscience over a fixed and harsh world. The idea of the senses feeding the brain to create a person's knowledge base became widely accepted during the eighteenth century. These issues were debated in influential intellectual salons in the latter half of that century, almost exclusively led by women.[iv] It was so widely accepted that the most popular book of that century, written by Thomas Paine and published in 1776, was called *Common Sense*. It was a political pamphlet that was steeped in the ideas of acquired sensory knowledge, and was hugely influential on the composition of the American Declaration of Independence some months later. In *Common Sense*, Paine laid out a clear case for a natural equality, rather than an innate inequality, among people. This could not

have been written without the Sensationalist idea that we are constructs of acquired sensory information.

The historical and philosophical ideas about knowledge, spirit-
uality, soul and mind remain. The splitting of human experience
and function into 'body, mind and soul' continues to permeate
most cultures. The common denominator of all these religious
and spiritual systems is that of implanted or external knowledge,
the third eye, a force beyond the individual. Neuroscience is fascinating to us not just because it helps us understand ourselves, for
which we have an insatiable curiosity, but also because it liberates
us with scissor-snip conclusiveness from third-eye phenomena.
Strangely, neuroscience has not yet permeated the culture of
psychiatry as one would expect. There is still a widespread perception that psychiatry, as a form of medicine, does not necessarily
involve the brain, but a hypothesized mind/spirit domain of the
human condition. Dualism for me, as a psychiatrist, is the enemy,
whether the dualism is that of the body–brain, brain–mind, body–
soul, reason–emotion. The divisions between these made-up
domains collapse when you realize that the world is conveyed to
you *only* through your senses and that we make sense of it all
through the pervasive connectivity of brain circuitry (a phrase
that I have borrowed from the physicist-neuroscientist Danielle
Bassett[4]).

The internalization of the world in a human being is transmitted through the Big 5 – sight, sound, touch, taste and smell – that
are being fed continually into memory networks. Sensation from
the world, such as touch or sight, allows you to learn different
shapes, forming relatively simple knowledge upon which more
complex information gets built. There is also the often-ignored but
constant feed of sensory information from the body to the brain
that gives rise to feelings from simple emotions to complex feeling
states. Sensation is the fundamental raw ingredient that feeds the
brain: the substrate upon which the pervasive connectivity in the
brain is based. Memory is, in its essence, the infinitely complex

neural representation of sensory information that has been carried to the brain.

Learning Through the Senses

The slow process of learning through the senses can be seen in how infants develop, and how we teach them about the sensate world. It can be difficult to appreciate so-called intuitive knowledge – which is really the automatic processing of learned knowledge – except when one thinks about how an infant develops through sensate experience. We love their innocence of not knowing things that we consider to be intuitive: 'When will Dada grow down to be my size?', the magical disappearing and reappearing of peek-a-boo, the untutored emotional responses. There are armies of neuro-scientists researching how babies learn through sight, sound, touch, smell and taste. A leading developmental psychologist puts it well: 'Babies know squat,' a modern iteration of the *tabula rasa* of Aristotle and Locke.[v]

A more startling example, following on from Molyneux's Question, is how newly sighted adults like Oliver Sacks' Virgil learn about the visual world. They have to be slowly exposed to images of the world to learn what these images represent – otherwise they will be overcome by a flood of visual sensory information that they are unable to process. Because of this, newly sighted individuals are kept in minimally visually stimulating environments and gradually exposed to the world of images. This is not difficult to understand once you grasp the idea that everything has to be learned through the senses. When we say that we *see* something we mean that we see an image in our brain that we interpret as being something. You see a Rubik's cube or you see a tennis ball and you're not confused – you don't have to touch them to decide which is which. You *know* the difference between a cube and a sphere. What you are really saying when you say that you see a Rubik's cube or a tennis ball is that you have learned that the image you are looking at is a Rubik's cube or a tennis ball. What we call a sense is also a memory: *seeing* is both the immediacy of the sight of the object and the identification of the image. This brings us back to Bergson's quotation from 1896 that I opened this chapter with: 'In truth, all sensation is already memory.'

Generally, perception is organized in line with the sensory information that is being fed to the brain. We *make sense* of whatever our senses feed into the brain, forming memory pathways and interpretative frameworks that are dynamic. The simultaneous experience of sensation and memory can be understood if you are listening to a piece of music familiar to you that you are unable to immediately identify. Trying to identify the tune involves sensory and memory circuitry. More sensory information is coming in to augment and stimulate further sensory–memory integration. Sometimes we abandon our attempts to remember, knowing that the conscious search will probably fail and that it will come to us later. This silent processing may lead to the identification of the tune some minutes later. Memory is not static: it is in a state of flux in a never-ending dance with sensation.

Try to imagine that your peripheral senses are intact, you have 20/20 vision and excellent hearing – neurologically intact, as we medics say – but you see and hear things that others do not. We are now moving into the domain of hallucinatory experiences, where subjective sensory experience is no longer a reliable transmitter of information from the world around you. Hallucinatory experiences may come from a misinterpretation of sensory signals, if someone has a fever, for example, or from a phantom limb in a below-knee amputee, where the signal seems to be coming from the part of the leg that has been amputated. In psychosis there may be no incoming sensation when sounds are heard that are experienced as coming *from the outside world* or images can be seen that are not apparent to others. Sensate experiences determine a person's understanding of the world. In 'The Yellow Wallpaper' the diarist deduces that there is a woman behind the wallpaper because she sees her in the moving patterns, feels her behind the wallpaper, smells the pungent odour of the wallpaper and can hear her groans. In 1925 Virginia Woolf, probably when psychotic and hallucinating, received the same 'rest cure' treatment as had Charlotte Perkins Gilman in 1884, from the same neurologist, Silas Weir Mitchell. Rest cure as previously outlined involved almost complete isolation and forced inactivity. To be put in a situation nearing sensory deprivation with the wild, uncontainable sensory experience of psychosis must have been torture and could only have exacerbated the psychosis.

Virginia Woolf killed herself in 1941 because she was unable to endure any longer the torture of psychosis. Charlotte Perkins Gilman also killed herself, in 1935, writing in her suicide note that she would rather die by her own hand than suffer a slow death from cancer. In the next chapter we will explore how we *make sense* of sensation and how we may not all share a *common sense* of the world.

3.   Making Sense

Our intelligence is the prolongation of our senses.

Henri Bergson (1859–1924)

Some years ago, I was at a dinner party when a conversation about psychics started. Everyone was giving their opinion, some more eagerly than others, as is always the case. My turn came, and my two cents' worth, a bit off-piste, was that while most commercial psychics were charlatans who exploited individuals in emotionally vulnerable situations, there could occasionally be individual psychics who have psychotic experiences themselves, and may attribute their hallucinatory voices to the dead person who is being 'communicated with'. A woman, who appeared to be not in the least eccentric except perhaps that she 'believed in' psychics, looked to me and said, '. . . if you're not psychotic, where do the voices come from?' I had already replied 'What voices?' before I realized that she was telling the rest of the dinner party that hearing voices, for her, was a normal experience. Voice-hearing is surprisingly common in the general population – 10 per cent of people will experience hearing voices at some point in their lives – and does not mean that the voice-hearer will go on to develop a psychotic disorder.[5,6]

My dinner guest was simply making sense of her sensate experiences. For the voice-hearer, auditory hallucinatory experience may be a voice in their head that seems to be separate from them, or, at the other end of the spectrum, the voices may be just as real as voices coming from the external world. The latter – real external auditory hallucinations – are common in individuals who have a diagnosis of schizophrenia. How they make sense of their voices and other hallucinatory experiences gives an insight into how dependent we are on sensory experience in how we *make sense* of the world.

Joseph is a young man who has a diagnosis of schizophrenia. This is the story of how he made sense of his auditory experiences.

Joseph had been advised for many years by his family, friends and the family doctor to see a psychiatrist. The event that finally brought him to my clinic occurred in the preceding week in his local grocery store. He overheard a conversation between two men. One told the other that he wanted 'him dead by tomorrow'. Joseph did not know to whom they referred but knew that someone was in danger, possibly himself. He went to the police, who were very helpful, telling him that they would investigate, and also suggesting that he should go to the ER in the local hospital as he seemed very stressed. From the ER he was referred to the psychiatry services . . . and hence, in his early twenties, he came to my clinic.

He was a fine, healthy-looking young man with slouchy adolescent deportment, obligatory headphones, a quiet smile and a soft manner. He was dressed in loose-fitting, trendy clothes and was at ease, unusually so, when communicating his story. Joseph had been bright in school, having a particular flair for maths, physics and IT. He began to realize in his mid-teens that he had special gifts, a sixth sense, that allowed him to understand the world in a more profound way than normal folk. This, combined with his IT knowledge, allowed him to gradually understand what most people were blind to: that reality, as most people perceive it, is 'a simulation'. Since realization of this certainty, his life had centred on trying to understand who, or whose forces, orchestrated the simulation. His guess was that the unreality shared by most people was being technologically manufactured by the simulators. He did not know who was controlling the simulation, or what 'real' life was as experienced by most people. He was neither in the real world that was controlling the simulation, nor the unreality that other people shared. In Joseph's world, everyone's reality was being implanted through this mechanism, but he was one of the few people who knew this. *He knew because he was able to hear voices from the 'outside' – the real reality that was under the radar.*

In the beginning, in early to mid-adolescence, he heard occasional dim voices that were difficult to understand. He could only make out bits and pieces of what they were mumbling when he concentrated hard on listening to them. Smoking cannabis cleared his mind of the clutter of the everyday, and he could hear the voices clearly. His friends told him that he talked rubbish when he smoked, and Joseph thought that this was ironic because he was the only one who understood what was really happening, and he did so with greater clarity when smoking weed. His family became more and more exasperated by his self-imposed isolation in his room, his dope-smoking, and his apparent laziness and lack of interest in his studies. Joseph did not feel uncomfortable being cut loose from the world and even felt creative in the complex paranoid puzzle. The influence of the simulator spread and demanded more of his attention as time went by. In the early days it leaked occasionally through the TV in his room, later spreading to his phone and then his headphones.

He went on to study computer science in college. These studies would facilitate his investigation of the origin of the outside force: it had to be transmitted, he reckoned, through some form of centralized digital hub. After graduation he decided to do a higher degree in cyber security. Joseph could now hear the voices all the time and 'in stereo'. He could hear them speaking when he was talking to others, competing with and often invalidating what present people were saying. When he heard the voices in a crowd, he wondered if the simulators might be among the strangers in the street. He would then wonder if there was a group of people who 'knew'. Sometimes he thought that the knowing others were signalling to him, the signals disguised as casual gestures, but he couldn't be sure. Another possibility was that the outside force had transiently entered the body of a stranger to communicate something to him. Walking out was a minefield of complex interpretations and paranoia, and he became more and more reclusive.

Doing the degree in cyber security would give him more skills to out-manoeuvre the outside forces and possibly allow him to 'disappear'. He began to make false identity documents and created several aliases, *multiple bank accounts and fake utility bills that together would make it difficult for the simulators to trace him. The Fraud Squad caught up with him, however, and, not wanting to draw attention to the simulators, he did not mount a defence. He served a four-month jail sentence and became further ostracized socially. I saw him about a year after his release from prison. He spent almost all the time now in his room deciphering the voices and the codes. The controllers had gained a further foothold, and he worried that they could now force him to do something repugnant. They sometimes threatened this, and sometimes told him that he should kill himself. On the rare occasions when he was out and about they now found him easily, speaking to him all the time through random strangers. Joseph would then find refuge in an internet café, where his signal could get lost among other competing signals.*

He had been conflicted about taking 'treatment' but wanted to rid himself of his burdensome insights. In the end he decided that he wanted to 'enter society again', even if it meant losing his awareness of the real reality. Joseph used an interesting phrase – 'I want my attention back' – as a reason for taking treatment. Over the first few weeks of anti-psychotic treatment the voices became less intrusive, and, as we increased the medication, they gradually faded over a period of months. The ideas that followed from his voice-hearing began to disassemble, leaving a strange empty wasteland of personal memory. He slowly became re-connected with his family and friends, and the stronger these connections became the further away he moved from the psychotic world of the simulators. Our occupational therapist introduced Joseph to a computer workshop that in time he helped to run before he moved on to successfully apply for IT jobs in the open market. Joseph now believes that his psychosis gave him entry to a world that does exist 'somewhere in cyberspace', but he is no longer interested in pursuing the isolating paranoid labyrinth.

Joseph's story shows how one makes sense of the world through one's senses and how this becomes the basis for one's interpretation of the world and for one's memory. Joseph did not share in the *common* sense of the world. Joseph created a narrative from his sensate experiences, as we all do. His colonization by psychosis had made him a stranger in the world of common shared experience. Like Virgil, the man who gained sight at 55 years of age and who had to build up a memory system for visual information, Joseph had to construct a new framework within which he could interpret and memorize the world post-psychosis. He has a diagnosis of schizophrenia – a broad diagnosis – in which the core experiences are hallucinations, usually hearing voices, and delusions, or bizarre beliefs. Organized psychotic belief systems, or delusions, gradually develop over time: Joseph came to organize the information coming from the sensate world into plausible and superficially coherent systems. If looked at from the outside, Joseph's worldviews were 'crazy', but they had an internal logic that explained his experiences. His scheme of the world was really quite clever.

Interpreting Sensation

How do we make sense of what the senses feed to the brain? A basic rule is that nerves from a particular part of the body go to a particular part of the brain for interpretation. The brain is a gelatinous blob of uncooked-shrimp colour, with squiggly indentations. The brain blob is fitted snugly into the concave grooves of the interior surface of the skull, similar to the shrimp in its shell, or a highly ruched walnut in a shell. The inside of one half of the walnut shell is similar in principle to the inside of the skull. Like a walnut, the brain is visibly divided into two halves. We say that the brain has two hemispheres, but actually the whole brain is more like a hemisphere composed of two quarter-spheres, but we'll stick with the familiar convention of hemispheres.

The outside layer of the brain is called the cortex, and this is the destination for most of the nerves that come from the body. We think about the cortex as being divided into zones, like countries on a map, each country receiving neurons from a particular part of

Figure 1. The sensory path from the senses to their cortex

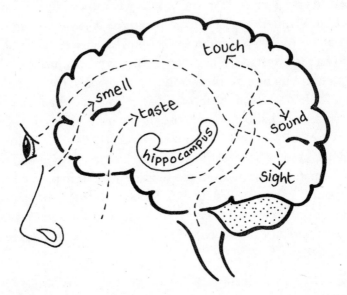

Illustration of the brain on the outside. The nerve tracts go to the outside of the brain, called the cortex. This demonstrates where the 'Big 5' – hearing (auditory), sight (visual), smell (olfactory), taste (gustatory) and touch (somatosensory) – sensations are mapped on the cortex.

the body. We call this representation of the body-in-brain 'brain mapping'. The Big 5 sense organs each have their own patch of cortex that is easy to delineate by the big fissures in the brain surface: the visual cortex (sight), the auditory cortex (sound), the olfactory cortex (smell), the gustatory cortex (taste) and the somatosensory cortex (touch) (see Figure 1).

Sensory neurons are nerve cells that enter the brain from all parts of the body, from the toes to the scalp. Neurons vary in design but in general have dendrites, hairy outgrowths, at one end that coalesce to a single strand of cytoplasm that ends in a nerve terminal (Figure 2). The nerve terminal is a blunted knob of cytoplasm containing

Figure 2. A typical neuron

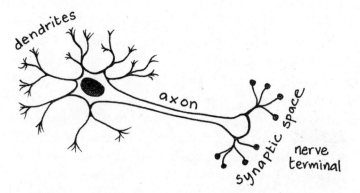

The electrical signal is conducted down the neuron from the dendrites to the nerve terminal. There, the signal changes the cell membrane microstructure and the bubbles of cell membrane containing neurotransmitters merge with the membrane surrounding the neuron, rupturing the bubble into a continuous surface with the outside, and releasing the contained neurotransmitter.

bubbles of neurotransmitters. The bubble of cell membrane containing the neurotransmitter, if stimulated by an electrical signal, merges with the membrane surrounding the cell, and the neurotransmitter spills out into the synapse. The synapse is the gap between one neuron and the next. The released neurotransmitter floats off and attaches to nearby receptors on adjacent dendrites, triggering an electrical wave in the post-synaptic neuron. This electrochemical signal is the basis of all brain activity.

The electrochemical energy that transmits the signal from one neuron to the next can be generated in ways other than through neurotransmitter release. Light may trigger an electrical signal in the retina, air-borne pheremones may light up a smell receptor in the nasal passages, food chemicals may trigger an electrical signal in taste buds on the tongue, sound waves may cause movement of the ear drum to generate a signal, or touching something may stimulate touch receptors under the skin to generate a small signal.

Touch Sensation

Back in the very early days of neuroscience, Molyneux and Locke could only imagine the sensory pathways of touch, from the fingers to the brain, and sight, from the eyes to the brain, that are now common knowledge. The sense organ for touch from the outer world is the skin. The signals from everything that we touch, or are touched by, arrive in the touch cortex – the visible bulge on the surface of the brain behind the big horizontal fissure (see Figure 1). Wilder Penfield (1891–1976) was a pioneering neurosurgeon, at a time when very few ventured into the brain, at the beginning of the twentieth century. He carried out experiments on conscious patients who were about to undergo brain surgery for epilepsy, and he found that pinpricking the touch cortex produced sensations of being touched in different body areas. Surprisingly, one does not feel any sensation on the touch cortex, because, although the brain interprets touch from on and in the body, there are no touch receptors on the brain itself. This counterintuitive fact, that the cortex cannot feel but feels the body, is a simple demonstration of the indivisibility of the brain–body complex. Over time, Penfield observed that pinpricking a similar location on the touch cortex of different patients produced sensations in the same location in the bodies of all the patients. He came to understand that the body was 'mapped' in a predictable way on the cortex of the brain.

In 1951 Penfield published a book with a map of the body as he found it to be represented on the touch cortex. He was the pioneer cartographer of the body-in-brain map, and his original sensory homunculus (meaning *small man*) has changed very little over the years. You can see from Figure 3 that the body is disproportionally represented in the touch cortex with the more sensitive zones, such as the mouth and lips, having a relatively large amount of nerves, and therefore cortical surface, compared to, for example, the legs, which are much bigger in corporal terms but are less sensitive because they have a lower concentration of touch receptors.

Figure 3. The somatosensory homunculus or touch cortex

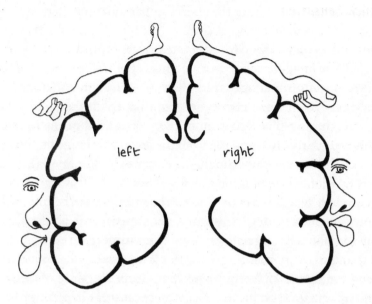

A vertical cross-section, from ear to ear, through the hemispheres of the brain at the level of the touch cortex. The touch cortex is located behind the big horizontal groove on the cortex. The mapping of the human sense of touch for general body parts can be seen. The more sensitive body areas have a disproportionately greater area of representation because more nerves leave the more sensitive parts of the body, like the lips and tongue.

As well as the outer cortex where nerves from the Big 5 arrive, there is another map for nerves that come from inside the body, delivering what is called 'interoceptive' sensation. The interoceptive cortex, called the insula, is in a fold of cortex not visible from the outer surface of the brain, rather like our feelings. The insula is very important because it interprets 'gut' feelings. We will look at this in a coming chapter, but just note for now that the way we experience the world is determined as much by the interoceptive as the exteroceptive inputs to the brain.

Penfield's work demonstrated, at a basic philosophical level, that feelings in the body can only be experienced through the brain.

Sometimes the brain can get the interpretation from its peripheral neurons wrong, such as in the case of a phantom limb. In phantom-limb pain, the pain sensation coming from the stump, say that of a below-knee amputation, is interpreted by the amputee as coming from the severed leg, now absent. This is because the *nerve* going down to the original foot, although no longer there, continues from the severed nerve in the knee stump up to the foot area of the sensory homunculus. The map of the foot is fixed in the sensory homunculus, that is, the nerve from the lower part of the leg always goes to a particular part of the sensory cortex. This nerve tract has been memorized in the touch cortex, and it will fire the 'lower limb' patch of cortex even if it has been amputated. This is an example of hard wiring – the anatomy of the nerve from the lower limb to the touch cortex is fixed. If, in a situation of reversed ana-tomical pathology, the cortex for the lower limb is damaged, for example following a stroke, there can be no experience of the leg, although the stroke survivor may have a perfectly formed limb. We experience the body only because of the brain.

The hard-wired sensory homunculus of Wilder Penfield's can now be seen in humans using the brain neuroimaging technique of magnetic resonance imaging (MRI).[i] Arno Villringer's group, only fifty years later, were able to touch parts of the hand to identify the areas of the touch cortex that would 'light up' using MRI.[7] The leap from Penfield's homunculus derived from open brain surgery to the visualization of the brain through neuroimaging is an example of how neuroimaging tools have opened up brain exploration.

Multimodal Sensation

The core principle of the mapping of the body in the brain is that every bit of the body outside the brain is represented in the brain. Within the brain it is all pervasively and uniquely interconnected, fashioned by the hard wiring of the genetically inherited brain and the soft wiring of experience that is grown by sensation, in a complex interplay of memory and learning. Soft wiring is how the brain wires to adapt and to learn. There is an ongoing organic interplay between hard and soft wiring. For example, non-sighted individuals read the world through touch. The anatomical basis for seeing through touch is that neurons from the touch cortex go, via cross-cortical connections on the surface of the brain, to the visual cortex. 'Touch' sensation then gets translated into 'sight' information. Individuals who are non-sighted from birth, compared to sighted people, have a smaller visual cortex but more cross-sensory cortical connections to the visual cortex from other cortical regions.[8] Non-sighted individuals do 'see', but the images are two-dimensional or flat: they do not have visual field depth. The way that non-sighted people 'see' tells us a lot about the abstract nature of sensation. Sensation is primarily about *discrimination* between incoming stimuli, and discriminatory mechanisms are common across all sensory interpretations. The brain learns to discriminate one shape from another by touch, one image from another by shape, and so on. The same with sound. Sounds are identified in relation to other sounds: think of musical scales and keys, or the different beats of music. Discrimination of one image or sound from another leads to pattern formation and hence recognition, the basis of memory. If the congenitally non-sighted individual is fortunate enough, like Virgil, to have sight restored, the visual cortex will become re-organized around input from external images, rather than from the touch cortex.[ii]

Another common example of cross-sensory memory is lip reading, and, without either sight or hearing, visual and auditory stimuli can be learned through touch and vibration. The cortical adaptation, with the shunting of sensory information around the cortical sensory map, shows how an individual's brain can adapt to memorizing and interpreting sensory information.[8] Brain plasticity – the ability of the brain to grow and adapt – has now been exploited for the non-sighted using sensory substitution devices that convert visual imagery, via a camera fixed on a person, to sound, so that the auditory cortex can learn to interpret the images caught on camera.[9] The language of the neuroscience of the senses has shifted from a sense-specific cortical model to a computational meta-modal organizational model reflecting the highly connected brain.

Sensation and Perception

In his book *Ways of Seeing*, and the BBC series that followed in 1972, John Berger (1926–2017) explored the concept of how we 'see' beyond the sensation of sight. He was an influential communicator on how the human brain interprets what is seen: *perception*. Berger, as an art critic, visual artist and writer, demonstrated how ways of seeing develop from our learned framework of the world, what he called 'perceptual constancy'. We see women and men in different ways, for example, because of deeply conditioned value systems. Perceptual constancy is necessary, or we would be in a state of constant re-evaluation of every stimulus, but it is also the basis for prejudices. Berger had first-hand experience of learned ways of seeing the world because he gradually lost his own sight owing to growing cataracts. Following cataract surgery his sight greatly improved.[10] He describes the 'visual renaissance' that he experienced post-op as if he was seeing everything for the first time. His visual memory circuits were being re-charged by light after many years of quietude, re-awakening his foundational visual memories. Vivid visual childhood memories returned: a white sheet of paper brought him back to his mother's kitchen and 'those whites on the table, on the sink, on the shelves'.

In *Over The Backyard Wall* the Irish writer Thomas Kilroy has written about a similar experience of visual rebirth following the surgical removal of cataracts. Kilroy visits his childhood memories by travelling in his imagination through the geography of his town of origin, Callan. This book was of particular interest to me because I was reared for some years in this small Irish midland town. I vicariously enjoyed his visual renaissance as he was describing the awakening of old visual memories. He writes that the 'actual origin of sight itself' is memory. Vision, like all sensation, cannot be separated from memory, and the two interweave to form perception.

To return to the psychic dinner guest who had benign auditory hallucinations, and to Joseph who had persistent intrusive hallucinations: what might be the sensory mechanism underlying auditory hallucinations? We don't know exactly, but we do know that the parts of the brain involved in hearing normal speech are firing when a person with psychosis is hearing voices.[11,12] You can imagine how, if the visual cortex was firing off, one might see wallpaper undulating; if the smell cortex was firing off, one might experience a rotting smell; if the taste cortex was firing off, one's food may taste strange, as if poisoned, perhaps. There is now evidence that the overall neural wiring of the sensory pathways in the brain that project to the cortex is different in schizophrenia.[13] This means that any problem that interferes with any part of the sensory pathways within the brain could potentially impact on sensory experiences. There are multiple reasons why the different sensory cortices may fire off in the absence of sensation coming in from the peripheral nerves: epilepsy, a brain tumor, local chemical stimulation from psychedelic drugs, miswiring, neurotransmitter malfunction. To put it pithily, as medical students are taught: a hallucination is the experience of a sensation in the absence of an external stimulus or, more correctly, a *matched* external stimulus.

Virginia Woolf suffered from bipolar mood disorder, and it is beyond doubt to me that her descriptions of the sensory experiences of Septimus, in *Mrs Dalloway*, were informed by her own experiences when psychotic and manic. In mania there can be a normal delivery of external sensation, but this is experienced in an exaggerated hyper-perceptive way, or may be misinterpreted if the person has psychotic mania. Sitting under trees in Regents Park, London, Septimus feels that the 'leaves were alive, trees were alive. And the trees had been connected by millions of fibres with his own body'. This passage ends with his conclusion that, 'All taken together meant the birth of a new religion.' We can feel the tremulous hyper-perception of Septimus, his fragmented understanding of the visual and auditory experiences, and then . . . his deluded interpretation: 'All taken together meant the birth of a new religion.' He had intact sensory delivery of sensation – the tree and leaves were real – but he did not make sense of it. Woolf's is one of the best descriptions that I have ever read of what in psychiatry we call a delusional perception, or a bizzare idea that arises from real externally derived sensations. In the end, Septimus, unable to endure the mental bombardment of psychosis, kills himself, fore-telling the death by planned suicide of Woolf herself.

Many people, including psychiatrists, argue that we should scrap the diagnosis of schizophrenia on the basis that psychotic experi-ences are common and that schizophrenia is a spectrum disorder. This, they argue, would have the added benefit of de-stigmatizing the disorder. For me, changing the word does not do it. The stigma is not present because of the word 'schizophrenia', and removing this word is an acknowledgement that it is indeed stigmatizing. Autism has become de-stigmatized, not because the word has been replaced, but because society has been educated. Individuals with autistic spec-trum disorder (ASD) are not seen as 'weird' any more; they are rightly seen as different/special. The insight and tolerance required to understand special, non-neurotypic, individuals raises us all to higher levels of kindness and generosity. But there is a very impor-tant caveat: there is a world of a difference between being on an autistic spectrum and having autism; or being on a psychosis spec-trum and having schizophrenia; or being obsessional and having OCD; and as for the casual remark that someone who is unpredict-able and emotionally unstable is 'bipolar' . . .! As I've mentioned, new knowledge awareness is usually overgeneralized initially and, in matters of the brain and behaviour, we tend to apply new informa-tion to ourselves first because we are so interested in understanding ourselves. Traits or tendencies do not make a disease, however, that can erode one's life, and this sort of self-diagnosis diminishes the sometimes devastating suffering experienced by those with full-blown mental illness. Hearing voices is not schizophrenia, any more than feeling sad is depression, or being highly organized to the point of rigidity is OCD, or having poor communication skills is autism, or being emotionally unpredictable is bipolar illness.

Interpretations of Strange Sensations

The distinction between delusional belief systems and the strange things that 'ordinary' people believe can be problematic. Some of the non-scientific marginal cultural belief systems, the third-eye phenomena, can seem quasi-psychotic, and some *are* frankly psychotic. It is noteworthy that some movies have a particular resonance and can provide frameworks of explanation for those with psychotic experiences. Many of the young men now attending our services with diagnoses of schizophrenia reference the mind-bending psychological thriller genre, typically *The Matrix, Inception* or *Memento*, to give a framework to their experiences of living in strange and fluid sensate states. *Inception*, in which the protagonist, Cobb, steals a person's thoughts while they are dreaming, had a profound influence on my patient Joseph. Cobb eventually trains himself to implant thoughts into the brain of another person while that person is sleeping. *Inception* demonstrated to Joseph how an alien idea could be implanted into a human brain and how, as Cobb says, 'once implanted it is impossible to combat'. Sometimes the movies are set in dystopian futures, bringing a dislocated sense of time to the sensory confusion – the cyberpunk genre that was led by Ridley Scott's original and brilliant *Blade Runner* is an example.

Each of us is a mixture of soft and hard neural wiring that develops from an initial state of nerve pathways laid down in fetal development and that grows with input from the world of experience. Sensory information becomes discriminated as memory becomes more elaborated through experience. This is the basis of perception and the perceptual constancy within which we automatically filter the world, giving each of us an individual and unique filter, our memory. In the uncut version of *Blade Runner*, Decker, disturbed by the human qualities of the replicants, wonders 'why a replicant would collect photos. Maybe they were like Rachael. They needed memories.'

4.   The Story of the Hippocampus

Long ago, in a place far away, a princess . . . the cosy composition of the traditional story is as familiar to us individually as it is culturally ubiquitous. We think that we intuit the time-place-person format because we have imbibed it as far back as stories have been recorded. Think of the format when you watch your next few movies: you are usually given the coordinates of time, place and person within the first few minutes. If not, you will be held watching until either you give up because it hasn't hooked you, or you will get a character, a location and/or time before you engage, and with a silent sigh you fall into the story. Although the format feels intuitive, it would be a mistake to attribute this to being steeped in it from the beginning of our lives.

The first thing that medical students and nurses are taught about mental state examination is what we call 'orientation in time, place and person'. This means knowing the approximate time of the day, the day of the week, the month and the year, and where and who you are. If a patient is not 'orientated by three', we know that something is seriously wrong with their brain function, whether lasting or temporary. We are actually, factually, neuronally wired to learn through the coordinates of time-place-person. The story of how we are wired to learn through these coordinates is the story of the hippocampus, and the matter of memory.

Try to imagine a way of being in the world without the story coordinates. What happens to an individual without a time-place-person format within which to process experience? Samuel Beckett is king of the exigent without a past, without memory, without identity. Many of Beckett's characters are amnesic and seem paralysed in an ever-present world. Vladimir and Estragon, the unfortunate tramps in *Waiting for Godot*, are trapped in the present, with no past and no future, waiting for someone – 'the saviour' – who may not exist and the audience knows will never arrive. The tramps are like two sides of a coin and seem unable to separate from each other or to derive any comfort from each other, a twosome without individualized personhood, anguishingly alone in a world without time and without place. They 'exist' outside the time-place-person format of the cosy story. They cannot remember the past (even the previous day), establish the present or project into the future, and go around in a blind circadian circuit seemingly waiting for death or salvation. Estragon's remark that there is 'no lack of void' encapsulates their disorientation. Beckett may have been writing about a world without God and filled with human cruelty and uncertainty, but he cleverly parsed down the experience of living to the time-place-person format that we live within. By removing the familiar coordinates, Beckett makes us feel how it is to live outside the orientation of memory. In the previous chapter we looked at sensory memory; now let's look at the elaboration of sensory information into the hippocampal memory format of time-place-person.

I witnessed, early in my training in psychiatry, what could happen to an individual in real life if they lost the capacity to form hippocampal memory. It was my first year of training in psychiatry in the late 1980s and I was working in St Patrick's Hospital in inner-city Dublin. St Patrick's was founded in 1746 by Jonathan Swift, a Protestant clergyman. Swift is best known for writing *Gulliver's Travels* and in his day was an infamous satirist. In Ireland he is better known for penning the outrageous satire *A Modest Proposal*, in which he laid out with deadpan irony a potential solution to the poverty that the Irish suffered under English rule – for the rich English to eat cherubic Irish babies. He wrote his own epitaph, referring to his posthumous self resting in his grave 'where savage indignation can no longer lacerate his heart'. Another 'savage indignation' that pained the living Swift was how the insane were ignored and abused by society. Swift had been a former governor of the Bedlam in London that became the Bethlem Royal where I treated Edith, and had brought his enlightened views about the humane treatment of the insane from the Bedlam to Dublin. Incidentally, Swift and Molyneux, of Molyneux's Question, lived during the same time in history and in the same place – a short walk from each other in Dublin – and were mutual admirers. They were driven by key ideals of the Enlightment – Molyneux that the brain was a human organ that processed knowledge from the human world, and Swift by an understanding that insanity was a disease of the human mind that needed humanitarian care. By the time I was working there, St Patrick's Hospital had little in common with Swift's founding principles of equality and charity, and had become a private hospital.

The Case of MM

MM had been referred for investigation of memory loss and personality change. She was a middle-aged, as I then saw it, woman in her early forties who was a homemaker with two children of early double-digit ages. I conducted the clinical assessment in an interview room facing the reception desk in the hospital foyer. I remember being in that room and MM sitting across the desk from me with an anxious, mistrustful face, frowning under her eyebrows at this strange young doctor. We decided that I would interview MM and her mother together because Mother was needed to provide MM's history.

Mother told me that MM had been behaving in an increasingly strange and uncharacteristic way, starting a few months previously. Apart from her almost complete memory loss, Mother noted that MM had become distant emotionally, even from her children. She sometimes did not recognize members of her family anymore, even her children, but this was not consistent. Her normal personality had been warm and she had been a successful mother and daughter. MM's parents had stepped in to take over the task of collecting the children from school, as she could no longer find her way to the school. The grandparents provided emotional support for the children, who were bewildered and upset by their mother's unpredictable responses. MM's husband was unable to cope with what seemed to him to be hostile behaviour from his wife. Each encounter and event that MM experienced seemed to be a new event not connected to anything previously, even if an event had happened only minutes before.

I particularly remember Mother saying to me that her daughter was always lost: if she left a room, even momentarily, and then returned, the room was new to her and she had to re-orientate herself. Testing this, I asked her to pause and look around the interview room and then accompanied her and Mother outside to the hospital foyer, closing the door to the room. I asked her again to pause and take a good look around the foyer and we turned around and returned to the room. The interview room was completely new again to her, and she was unable to identify it. Her disorientation had increased to the point where she could no longer be left alone. She was in a state of constant fear, and everything that happened seemed to be continuously new and frightening for her. MM, in spite of her crippling memory loss, remained capable of doing complicated motor tasks such as cooking and writing. When I talked to her, although she was bewildered and terrified she was unable to pinpoint her problem. She was able to see and identify objects. She was able to hear words, understand language and respond coherently. She could taste and smell. One of the problems that she did understand about her predicament was her complete loss of memory for location – she felt an overwhelming sense of being lost all the time.

MM struck me as being absent in all ways except that she was sitting in a chair at the other side of the desk. She had a distant and distrustful affect. ('Affect' is the word clinicians use to describe the impression that the emotional state of someone makes on you.) I knew that she did not know who I was, and the interview was strange because our rapport did not warm or change in any way as the consultation progressed. I had a sinking feeling as we were talking that day that MM would not come back and that her condition was beyond the remit of medicine to repair. I felt that I was simply an observer of some catastrophic brain event.

MM had what seemed like total short-term memory loss. I had never seen someone previously, nor have since, who had such a complete absence of short-term memory, with otherwise normal mental function, and with full sensory awareness. It is difficult to imagine. We are familiar with dementia, when memory function

deteriorates in tandem with other brain functions, such as understanding speech or speaking coherently. Individuals who lose their
memory usually also lose their ability to do tasks such as cooking or
driving a car. MM had excellent sight, hearing and touch sensation.
Unlike the blind man of Molyneux's Question, who was unable to
make sense of imagery, MM was able to make sense of images and
indeed all other sensations, but she lacked the capacity to make
sense at another level of memory – time, place and person – which
we will henceforth call 'event' memory. Event memory involves a
bringing together of disparate sensory information in the dynamic
living world. It is memory for what *happens*. MM could not commit
to memory what was happening around her or to her. With this she
had lost her ability to form biographical memory.

You may wonder why MM had been admitted to a psychiatric
hospital. I had been instructed to admit MM for investigation of a
possible 'hysterical amnesia' because her brain function, apart from
memory, appeared to be intact and functioning. Hysterical amnesia,
now called dissociative amnesia, was, and is, taught to trainee
psychiatrists as being a state of dissociation in which memory
abruptly stops working while other mental functions remain intact.
This leaves the patient with a circumscribed memory loss but with
otherwise normal brain function. Trauma is theorized to be the
precipitant of the dissociation: that is, someone has been so traumatized that they are unable to commit the event to memory, with
a resultant 'blockage' of general memory function. The impaired
memory, it is hypothesized, protects these individuals from the
potentially overwhelming emotional experience of recalling the
event or the memory. The treatment for dissociative amnesia is to
unveil the trauma and guide the patient through the experience of
the trauma, thereby restoring the flow of memory function. I have
not seen a case of hysterical/dissociative amnesia in thirty-six years
of clinical practice, and it is generally seen as an outdated diagnosis,
but the underlying explanations put forward historically for this
alleged state of mind are important in understanding some of the
lingering misconceptions about psychiatry.

Hysterical Amnesia

Hysteria was a very common diagnosis given to women in the late nineteenth and early twentieth centuries. Freud's account of his famous patient Anna O in *Studies on Hysteria* (*Studien Über Hysterie*, 1895), is a richly layered story that deserves revisiting because it contains many of the fallacious tenets of Freudian theory. Anna O was a pseudonym for a strong, intelligent feminist who became a patient of Freud through his mentor, the neurologist Josef Breuer. She was being treated by Breuer because she developed strange states where she appeared to lose awareness, to repeat words or gestures and experience visual and auditory hallucinations. Breuer noticed that her symptoms appeared to change when she talked about them – what we would nowadays call inconsistent accounts or hysterical elaborations – and he went on to try out what became known as the 'talking cure'. The talking cure was the basis of what Freud later elaborated into psychoanalysis.

Freud believed that Anna O's states, and those of most of his patients, were caused by some form of repressed memory, usually of a sexual nature. His treatment – analysis of the psyche, or 'psycho-analysis' – was to facilitate the patient to freely associate, or to talk in a non-directive way. Free association was theorized to lead to revelations about the individual and ultimately to reveal the trauma. Contrary to Freud's claim in his written case history that Anna O made a full recovery following her so-called talking cure, she went on to be hospitalized many times. There has been much speculation since about Anna O's diagnosis. Descriptions of her abnormal states sound similar to temporal lobe epilepsy, or could have been accounted for by tubercular meningitis (TB in the membrane surrounding the brain), or by chloral hydrate or morphine addiction or withdrawal. There is now a list of speculated causes of Anna O's symptoms that were not recognized in those days.[14] Breuer, not in denial about Anna O's outcome, became increasingly distanced from Freud.

Breuer's divergence from what he saw as Freud's increasingly narrow ideas about talking cures, and over-attribution of the causes of neurosis to infant and child sexuality, reflects what has happened since in the discipline of psychiatry. The talking therapies now commonly used thankfully have little in common with the Freudian practice of non-directive free association therapy. Therapy, most commonly some form of cognitive-behaviour therapy, is now goal-directed and is emotionally contained. Nor is therapy based in Freudian theory. It is remarkable that the bizarre idea that penis envy is the basis for some neuroses in females was taught to trainee psychiatrists like me, with no sense of irony, only thirty years ago.[i] The idea that girls are sexually attracted to their fathers is repulsive to us now, and can be seen as providing a *bona fide* rationale for the rampant child sexual abuse of the time. (We will look at this in greater detail in a future chapter.) Psychotherapy these days is much closer to Breuer's solution-focused original experiment with Anna O than to Freud's undirected free-association time-unlimited psychoanalysis.

What characterizes many of the famous historical cases of hysteria is an intense relationship between the therapist and the patient, one of whom has an emotional, the other a professional, investment in the diagnosis. Anna O saw Breuer for two hours daily for some months, and these encounters were emotionally intense. Later, during one of her hospitalizations, another doctor fell in love with her. Some patients, and some psychiatrists, enjoy the melodrama of hysteria, whether this drama is so-called psychogenic amnesia or, even more dramatically, multiple personality disorder (called dissociative personality disorder in DSM 5*). I had one patient in recent years who was arousing great interest among the younger staff because she was 'dissociating' into three different personalities with different names, different genders and different characteristics. The younger staff were fascinated, watching and waiting for the next character

---

*The Diagnostic and Statistical Manual of Mental Disorders, 5th edition, is an internationally recognized diagnostic guide for psychiatric disorders. The diagnoses are based on symptom check-lists and not on subjective impressions, facilitating consistent diagnoses across the US and the world. There are three clusters of personality disorders: Cluster A (odd, eccentric), Cluster B (antisocial, histrionic, borderline), and Cluster C (anxious, fearful).

episode to emerge. I instructed that the patient be escorted to and observed in a private room when this occurred, and was only to be addressed, and responded to, when she became herself again. Her dissociative episodes petered out and she attended our psychologist, not to investigate her multiple personalities but to try to address her real and substantial problems. The attraction to hysteria can also be seen in the voyeurism that these cases arouse outside of medicine.[ii]

Hysteria, as a diagnosis, is now considered to be defunct, but the idea that a neurological impairment, usually a sensory or motor impairment, or memory loss, can have a 'psychological' or 'non-organic' cause remains accepted within clinical practice. The clinical nomenclature and the literature in this area is a minefield of unhelpful abstractions and terms that shuttle between neurology and psychiatry, but underlying all of it is the implication that some human experiences are 'psychological' and some are 'organic'. In the real organic life of the human, brain function and matter are indistinguishable, because every experience in the brain, whether a normal or an abnormal one, is based in matter and how that matter functions. Until the 'Decade of the Brain' in the 1990s, the discipline of psychiatry was considered by most people, including practitioners, to be in the domain of the intangible 'mind', while neurology was in the camp of the 'organic' brain.[iii] Neuroscience is quietly transcending mind–brain dualism with new insights about brain function, without having to argue the toss about brain versus mind. Clinical medicine understandably lags behind cutting-edge neuroscience, and the concept of whole-brain indivisibility of function has not fully permeated all medical disciplines.

Let's return to MM, who was admitted for investigation of hysterical amnesia. I could not elicit any trauma in her past, yet she presented to me as a very sick person who was terrified and bewildered. At this time, neuroimaging was in its infancy and was a very scarce resource clinically. MM did get a brain scan in the following weeks that demonstrated a large tumor at the centre of her brain and she was transferred to oncology for ongoing care. There were few details in the scan report, except that the hippocampus, on both right and left, was not visible. The tumor was inaccessible to surgery and she died shortly thereafter. MM was typical of many patients who were investigated for, and sometimes diagnosed with, hysteria, in the times before there were good investigative neuroimaging tools. The historical literature is rich with such cases. The case of a woman admitted to the Maudsley hospital in London in the 1950s and diagnosed with hysteria following input from the most acclaimed psychiatrists and neurologists of the time, and who died two years later from a brain tumor, is an infamous example.[15] I never saw MM again, but the tragic sense of her being an empty shell, a daughter lost to her mother, a wife lost to her husband, a mother lost to her children, and, most of all, a person lost to herself, has stayed with me. With her total event-memory loss she seemed to have lost her personhood.

MM taught me that without a hippocampus each one of us would wander about like Estragon and Vladimir in a timeless, disorientated state without a memory of passing events and unable to contemplate the future. Importantly, unlike Beckett's tragi-comical creations, MM was intensely distressed. She also taught me that you cannot make a past without first making a present. The hippocampus makes a present because it integrates the sensory inputs from the cortex into a story of the present. We will look at how the hippocampus makes time-place-person information – the basis for event memory – in coming chapters. We need to start with how the hippocampus processes the raw sensory information that was available to MM, to create something more integrated – a perception of the continuous present – that was absent in MM.

The Hippocampus

Understanding the anatomy of the hippocampus is important to understanding the flow of sensory information from the outside world, through the sense organs, to the cortex on the outside of the brain, and on to the hippocampal hub at the centre of the brain. The

Figure 4. The hippocampus

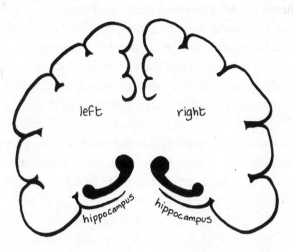

A vertical cross section through the brain showing the hippocampus at the curled-in rim of the outer cortex.

hippocampus is snugly tucked into the bottom rim of the cortex; think of a closed button mushroom cut lengthwise into two halves and look at the cut surface. The mushroom-coloured cap is the cortex and the hippocampus is the curled-in dark brown part of the mushroom where the cap meets the stem (see Figure 4). We have a hippocampus (plural, hippocampi) on each side of the brain – the brain being a mirrored structure – and while the right and the left hippocampus serve somewhat different memory functions they have a common mechanism of action. The name hippocampus derives from a Latin word meaning 'seahorse', referring to its shape. It has a big head, with chin tucked in and the body gradually tapering to a tail. It lies head-first in a front-to-back orientation of the brain.

Sensory memory, distributed around various cortical areas, is connected to the hippocampus through the neurons that converge on to the hippocampus from the cortex, like the gills from the cap of the mushroom that attach to the curled-in part (see Figure 5). Once the signals arrive from the cortex, they are processed through the cell-layers of the hippocampus that create new connections among the hippocampal cells. The signals cause hippocampal neurons to connect together, and the newly connected-up hippocampal neurons are essentially 'memory codes' of the nerve signals from the sensory cortices.

Before looking at the hippocampal coding processes in more detail, I'd like to tell the story of how neuroscience learned so much about hippocampal function from one man, Henry Molaison.

Figure 5. The sensory-memory path

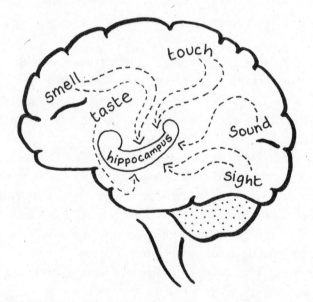

A vertical cross-section through the brain, cut in the middle in a profile orientation. Sensory information is transmitted through neuronal pathways from the sensory cortices to the hippocampus.

HM

The reason we know that the hippocampus is central to memory function in humans is largely due to the life of Henry Molaison (HM), the best-known case in the neuroscience of memory, whose history was written up in a landmark paper in 1957.[16] His clinical profile was very similar to MM's, although the cause of the hippocampal damage was different. HM fell off his bicycle when he was seven years old and damaged his hippocampus. The torn brain tissue led to the formation of scar tissue because brain tissue, like all body tissue, tends to scar during the healing process. Epilepsy, which HM subsequently developed, is often caused by scar tissue in the hippocampus: signals are blocked by the scar tissue and the electrical energy builds up, leading to an uncontrolled spreading of electrical signals in brain circuits. The brain is a mega-network of different circuits in which the hippo-campus is a central hub, and, if the current flow is dysregulated here, the whole brain can misfire. This can cause states in which all of the brain is misfiring simultaneously and nothing is being processed normally. In such circumstances, the unfortunate per-son will lose consciousness and fall, experiencing 'tonic–clonic seizures' in which the body muscles contract and relax uncon-trollably. The seizure, if not controlled, will over time cause further neural tissue damage.

Eventually, with doctors unable to control brain firing using even the strongest of anti-epileptic medication, HM had both his right and left hippocampus removed. In 1957, HM's surgery was a pioneering treatment, and removal of his two hippocampi did lead to a great improvement in his epilepsy. The unexpected and tragic consequence of this procedure, as you have probably predicted, was that HM suffered from profound memory loss for the rest of his life. Neurosurgeons now remove only one hippocampus because of the consequences of removing both, not known before the HM case. Following surgery, HM was unable to commit any event to memory. Every day brought with it a new world – a day preceded by no other days – of new places and new people. The house that he lived in was as unfamiliar to him from the first day of his post-surgery life as it was when he died fifty years later. Nothing was connected to past experiences, whether the past experience was two minutes previously or two decades previously. There was no past and no future, only an endless disjointed present – a staccato 'now'. HM, similar to MM, was alert, verbally fluent and had full motor function, but was still unable to recall exchanges in conversation beyond one or two sentences.

HM was studied in great detail following his bilateral hippocampectomy, particularly by a meticulous neuropsychologist called Brenda Milner, until his death in 2008, aged 82 years. Milner's mission was to try to find the reason why some memory function was intact and functioning in HM – even complicated sensory-motor tasks such as word-recognition and speech were still working – but he could somehow not put it all together outside of the moment to make *event* memory. Brenda Milner discovered that much of the memory that we use in routine daily tasks is stored in the cortex and may not involve using the memory factory of the hippocampal circuitry. This explains why HM could see, hear, touch, walk, cycle, converse – his cortex was intact – but the putting together of the 'who, where and when' to make an event memory was not possible. HM's visual cortex had already learned to understand the world by sight, so this information remained safely stored. Similarly, sound and smell, motor and language skills were all functioning. What he was missing was the past context, and any possible future context. The difference between memory stored in the cortex and event memory, which is processed in the hippocampus, can be seen in some rare cases of babies and children with severely impaired hippocampal function.[17] They can learn facts and figures, and language, and can even operate at an average academic level, but they do not make biographical or event memory.

Cells that Fire Together Wire Together

How hippocampal neurons make memories is the subject matter of a gargantuan literature and is one of the key questions at the hub of memory research. The neuroscience of memory is founded on the groundbreaking science of Donald Hebb (1904–85), who gave the world the catchphrase that sums up the neurophysiological process of memory: *cells that fire together wire together*. Hebb was a Canadian psychologist who worked with the amazingly creative group surrounding Wilder Penfield, who mapped the sensory homunculus. In 1949, in his book *The Organization of Behaviour*, Hebb described his theory of how neurons make memories, and how these memories go on to help organize brain function. He hypothesized that bunches of firing neurons connect up together to form cell assemblies that become wired together. The cells wire together through the formation of connecting dendrites made from the electrochemical energy of the nerve signal. The connected-up cell assembly subsequently fires as a single unit so that, if any of the constituent neurons of the cell assembly are then stimulated, all of the neurons will fire. This cell assembly represents a memory. To put this simply, a memory is represented in a neural code that consists of cells that have been wired together to fire as a single unit.

The inter-neuronal connections in the cell assembly could be, Hebb hypothesized, solidified by the physical growth of dendrites between the neurons, creating a more permanent memory, or alternatively they could fade away. The 'Hebbian' model of dendritic growth, through the firing and the subsequent increased connection of adjacent neurons, is now accepted as the cellular basis of memory. The dendrites are all-important in this process, because they transmit the nerve signal from one neuron to the next. Greater dendritic growth means increased connectivity among neurons. Dendrites grow in a quite beautiful way, called 'arborizing' (from the Latin *arbor*, meaning 'tree'), because the dendritic fibres resemble the branching of a tree. Neurons can have up to 15,000 dendritic outgrowths and, you may remember, there are 68 billion neurons in the human brain; consider, then, the astronomical possibilities of connectivity among them through dendritic arborization and new synaptic formation. It is not magic, but it almost seems to be so because there are virtually infinite possibilities.

The key process in forming even a short-term memory is that the cells have to *fire together* for long enough to become *wired together*. The firing together forms a transient memory and the wiring a more permanent memory. The process of strengthening the encoded cell assembly is called *consolidation*. Information is entering the brain constantly during the waking state, but most of it is not solidified – it simply fades because it has no relevance. On a molecular level, the formation of a solidly connected memory from an ignited cell assembly depends on the many factors that influence the strength of incoming signals. If the strength of the signal is at a key threshold, the neuron will make the dendritic proteins and the memory will become more permanent. If the signal is poor, the cell assembly firing will fade away, and there will be no wiring. Cells need energy to grow the dendrites and the energy comes from the electrical activity in the neurons – the more firing, the more wiring.

The Hebbian process of converting electrochemical energy from firing neurons to form proteins that constitute dendrites is a neat example of how, in the brain, energy is converted to matter. Hebb, like all great discoverers, meticulously observed his measures and faithfully recorded these observations in spite of the fact that he was not able to prove the theory that followed from these observations. My favourite thing about Hebb is that he believed that a theory, rather than being pitched against another theory, should be used to provoke thought and to guide research. One has to sometimes struggle through an enormous and theoretically framed literature in psychology to come to an understanding of what lies beneath the theories and counter-theories. Outsiders often have difficulty teasing apart juxtaposed theories that paradoxically can seem quite similar. And they *are* similar, because new theories emerge from pre-existing theories. Hebb did not cast his theories primarily in counter-position to existing frameworks but used this knowledge to move forward in terms of understanding his observations.

Hippocampal Plasticity and the Organization of Memory

There are a limited number of hippocampal neurons, and they are in a constant state of assembly, disassembly and re-assembly as we negotiate the present sensate world and consolidate, or not, a memory assembly. Hippocampal neurons have to be especially adaptable to allow continuous synaptic growth and re-modelling. The ability of a physiological system to change or re-model is called 'plasticity'. The hippocampus is intrinsically plastic and can actually be seen to 'grow' as a unit in some situations of intense memory-formation. A striking example of hippocampal growth following learning is the famous London taxi-driver study, in which the right hippocampus was found to be measurably bigger following two years of intensive learning of routes.[18] My 'sticky' memory of this study comes to mind every time I step into a London taxi. Memory deterioration, on the other hand, is part of normal ageing and can be visualized on MRI in the reducing size of the hippocampus as the brain ages.

We now know that depression is associated with a smaller left hippocampal volume and that this reduction is greater if depression is recurrent or has been long-lasting. You can now see that there appears to be a laterality effect (in which one side of the brain is more involved in a particular function than the other) for different types of memory: it is the right hippocampus that changes over time in the taxi study, while the left hippocampus is generally found to be smaller in depression. This is because the right hippocampus is more important for place memory, and the left for biographical memory. Given this, it is not surprising that depressed individuals typically have poor memory function, and there is usually patchy, sometimes lost, biographical memory for the duration of the depression.[19] A recently published study from my research group pinpoints that a specific area in the left hippocampus, where the 'coding' of cell assemblies occurs, is reduced in size in depression.[20] [iv] In this study we found that those who were having a first episode of depression did not have changes in their hippocampus, while those who had been depressed for longer periods of time had hippocampal changes. The happier news is that memory function improves once depression is treated.

Cortical Memory

Does all the memory remain in the hippocampus? No, because, as we have seen, there are a limited number of neurons in the hippocampus, and they need to be recycled to make new memories. So where does all the hippocampal memory go? The simple answer is that there is a constant dialogue between the hippocampus and the cortex, and much memory is ultimately stored in the cortex. Neurons in the cortex, in contrast to those in the hippocampus, are more difficult to change or rearrange – they are less plastic. This means that memory maps laid down by the multiplicity of interwoven cell assemblies in the cortex are relatively resistant to change and therefore to damage. Having the two systems of memory, one that is rapid and plastic, and one that is slower and more stable, means that we can continue to learn and adapt to change within a relatively stable system of knowledge. This does not mean that cortical memory is static, far from it. The cortical 'www' is in a state of ongoing interaction with the plastic hippocampus throughout one's life.[21]

The storage of event memory, of biographical memory, takes place in the neuronal dynamic between the hippocampus and a cortical area in the front of the brain, and in particular an area called the prefrontal cortex.[22] The prefrontal area is situated above the eyes, and it lights up when a person is being scanned while consciously recollecting a personal memory.[23] The situation is that the hippocampus seems to be involved in the laying down of event memory, and may also be involved in the recall of events from the past.[v] Although HM could not memorize his own life from the time of his surgery onwards, he could remember events from his childhood for up to about three years prior to the removal of his hippocampi. This is how it was first learned that although biographical memory is made in the hippocampus it is not for ever stored there. It was suspected that the three-year period of biographical amnesia prior to HM's surgery may have reflected the time that it took to process and transfer biographical memory from the relative instability of hippocampal cell assemblies to the more consolidated connections in the prefrontal cortex.[24] We now know that, as memories age, they diffuse up to the cortex from the hippocampus, and that it can take months or years for this process to occur. Neuroscientists can now see that the hippocampus primarily lights up when recent events are being recalled, whereas the prefrontal areas are engaged with the recall of more remote events.[25] HM had an intact prefrontal cortex and was able to access biographical memories that had woven their way to this higher part of the brain. 'Higher' is a word conventionally used when referring to parts of the brain, like the prefrontal cortex, that mediate complex functions like biographical memory. The hippocampal-prefrontal circuit is the main neuronal highway that processes your personal history over a lifetime.[24] The prefrontal cortex network is the storyteller, gathering information from all over the brain in 'working memory' to tell the story.

We have already seen how sensory memory – sight, sound, smell, taste and touch – is primarily organized in specific areas of the cortex. The visual cortex develops and memorizes images as a child develops, or this can happen, rarely, in adulthood when a newly sighted adult like Virgil emerges into a visual world. The visual cortex retains the image memory and can operate superficially as a hippocampal-independent system. This would be consistent with the experiences of HM and MM, who, without a hippocampus, had intact sensory knowledge. But this is an over-simplification and one that misses the experiential wonder of sensory memory. Remember John Berger's experience of a 'visual renaissance' following cataract surgery and the restoration of his sight: when he saw a white sheet of paper and was suddenly back in his mother's kitchen. In Berger's vignette the visual cortex was stimulating deep biographical memory. Visual art challenges automatic sensory interpretation – perceptual constancy – and brings us into Berger's world of breaking down perceptions.

The original, and probably still best, inroad to peeking into biographical hippocampal recall is the experimental method of stimulation of hippocampal cells in conscious patients during pre-operative epilepsy surgery. Oliver Sacks, in his famous *The Man Who Mistook His Wife For a Hat*, described the original shock when Wilder Penfield stimulated the hippocampus. Stimulation

> called forth intensely vivid hallucinations of tunes, people, scenes, which would be experienced, live, as compellingly real, in spite of the prosaic atmosphere of the operating room . . . such epileptic hallucinations are never phantasies: they are memories, and memories of the most precise and vivid kind, accompanied by the emotions which accompanied the original experience.

Going Cortical

Much of the process of daily memory 'going cortical' seems to occur during sleep. The effects of sleep on memory were first described by Hermann Ebbinghaus (1850–1909) in his seminal work on memory *Über das Gedächtnis* (*On Memory*), in 1885. Ebbinghaus examined his own pattern of memorizing and noted that new information was more likely to be memorized if presented before sleep than during the day. It became apparent in the sleep

and memory studies following this discovery that sleep deprivation impaired memory acquisition.[26] One of the reasons, it seems,
for the positive effects of sleep on mnemonic function is the electrical activity that occurs in the brain during sleep. Electrical
activity recorded from the scalp during rapid eye movement
(REM) sleep resembles the firing activity that wires hippocampal
cell assemblies. These rapid electrical brain cycles during sleep
represent the zapping of the cortex with the daily load of newly
formed memories from the hippocampus.[27] The 'off-line' memory
consolidation from the hippocampus to the cortex during sleep
has now been seen in mice.[28] The cortex zaps the hippocampus
during the day and the hippocampus zaps the cortex during sleep.
Dreams occur during REM and may be prophetic because present
events from the hippocampus, when going cortical during sleep,
may re-activate associated cortical memories laid down in the past,
releasing a glimpse, sometimes a disturbing one, of what has happened previously in similar circumstances . . . and therefore may
happen again.

In Beckett's novel *The Unnamable* (1949), the narrator has no
identity – there is only the voice, a disembodied voice, a stream of
words, that reaches an existential crisis: '. . . you must say words as
long as there are any until they find me, until they say me . . .
perhaps they have carried me to the threshold of my story . . .' Each
person's identity is a story, and, if there is no story, there is really no
self, more precisely no continuous sense of self – one is unnamable.[vi] *Unnamable*, like Vladimir and Estragon in *Waiting for Godot*,
gives us a disturbing visceral inkling of an existence without a past
or a future, an interminable sense of a disjointed, disorientated,
depersonalized existential present . . . a staccato 'now'. Beckett's
characters provide theatrical creations of the harrowing loss of self
that I imagine MM experienced: the ultimate existential crisis that
the hippocampus keeps at bay.

The often-quoted lines that form the conclusion in *The Unnamable* –
'You must go on. I can't go on. I'll go on.' – will always move us as a
profound expression of the sometimes unbearable, universal human

condition that must be borne. Being in a world, with an identity, even a disembodied voice created from language, compels one to go on. The hippocampus will take whatever is presented from the cortical world of your sensation and convert it through the cortex into your human story.

5.   The Sixth Sense: The Hidden Cortex

The smell of cut grass evokes for many of us the memory of a child-hood summer or many happy summers merged, siblings and first cousins running on lawns at family gatherings. The smell of potato clay turning to dust . . . the small grocery shop in Main Street, Portarlington; the sawdust and warm resin smell of newly cut wood . . . a Saturday morning in the hardware shop with my father; the heavy sour smell of a creamery . . . getting the country butter on Friday in the co-op in Green Street, Callan. In these memories we can often see ourselves as we were then. Marcel Proust's description of the immediacy and purity of the memory associated with the taste and smell of a madeleine, a small sponge cake, is famous, and perhaps over-referenced, but as a classic it is worth quoting again.

> But when from a long-distant past nothing subsists, after the people are dead, after the things are broken and scattered, taste and smell alone, more fragile but more enduring, more immaterial, more persistent, more faithful, remain poised for a long time, like souls, remembering, waiting, hoping, amid the ruins of all the rest; and bear unflinchingly, in the tiny and almost impalpable drop of their essence, the vast structure of recollection.*

Proust describes what we have all probably experienced at some time in our lives. We taste or smell something and we immediately *feel* the emotion associated with the memory of the taste or smell.

Smell and taste are interpreted in overlapping cortical areas, but smell is the more immediate trigger for emotional memories. The

---

* Marcel Proust, *In Search of Lost Time*, 1, *Swann's Way*, trans. C. K. Scott Moncrieff and T. Kilmartin (Vintage Books, 1996).

experience of a vivid emotional memory being triggered by a smell, wrapping you in your own mystery, is known as the Proustian effect. We, each of us, experience our own personal Proustian moments, and literature is full of arresting Proustian memories. John Banville has written of one such experience: 'lupins are for me what the madeleine was to Proust.' When he catches the smell of a lupin, 'time falls away and I am a child again.' He hears 'the sound of the sea', feels 'the sting of salt on sunburnt skin', tastes 'banana sandwiches', and smells 'that mingled smell of crushed grass, seaweed, nightsoil and cows . . .'* Smell is the most mysterious, soulful and delicate and the most intrinsically emotional of our senses, but it is not *immaterial*. In this chapter we will journey through this material soul, using smell to look at how emotion, what I will call the sixth sense, is woven into sensation and biographical memory. This will involve an exploration of interoceptive sensation, body feelings and their interpretation in the hidden emotional cortex, the insula – 'the rag and bone shop of the heart', in Yeats's memorable phrase.†

During the days of my first consultant post in Addenbrooke's Hospital in Cambridge, I had a Proustian visceral experience that I will never forget, which I will now recount.

The Cambridge Lovage Story

It was the warm sunny summer of 1995, and I was in the early months of my first pregnancy. We had bought a 300-year-old house with a garden so big that the back boundary was not visible to the eye from the kitchen door. If, that summer, you had followed the parched lawn down past the big oak tree and then the orchard, you would have entered an undergrowth with a thick smell of earth compost mixed with hot river vapour, before seeing the river that marked the back boundary. The hot air, as stagnant as the shrunken

---

* John Banville, 'Lupins and Moth-laden Nights in Rosslare', in *Possessed of a Past* (Pan Macmillan, 2012), p. 403.
† W. B. Yeats, 'The Circus Animals' Desertion'.

river, had trapped spinning clouds of small river insects. The mesmeric unending sounds of the invisible crickets, possibly grasshoppers, extended into mellow fermata as if sound was also trapped. I found myself in a mood similar to the weather, suspended and absorbed by nature, and I spent a lot of my time that summer in the herb garden, laid out among flagstones, by the back door. Most of the herbs had been planted by previous owners, and I spent weekends over months restoring it – weeding out ivy, mint and lemon verbena and cutting back the lavenders and thymes that had gone to wood. I was also learning to use the garden herbs in cooking.

One day during this lovely summer, I picked a bunch of herbs to flavour a salad. Although I had daily morning sickness, I was always well by evening, except for that night, when the nausea returned with a vengeance. The next morning I was more than usually sick and remember having to lie down on the wiry carpet tiles in my office between consultations to close my eyes for one or two minutes before facing the next anxiously waiting patient. Unable to identify what had made me ill, I felt that it was something in the green salad that had brought it on and decided that greens were out of my diet for the remainder of the pregnancy. Some days later while working in the herb garden I bent down and, inadvertently smelling a plant, felt a jolt of nausea. Looking around to identify the source of the smell I saw a tall, luscious green plant with yellow clusters of flowers that I instantly identified as the culprit in my sickness of the previous week. It was lovage, an ancient plant grown in convents and monasteries around Europe, used then extensively in cooking and in herbal remedies. It was puzzling that it had made me sick.

I was so certain about the identification of the smell that I left the garden to find my partner, Ivar, and tell him. He was also intrigued and interestingly did not question that I had identified the cause of the sickness. I didn't understand though . . . lovage was not harmful . . . I must have misidentified the plant or have misremembered what I had read about it. I consulted my herb books only to discover, in a hidden-away paragraph, that lovage should not be consumed during pregnancy. In olden times lovage was used as a herbal abortifacient. Large quantities had been traditionally used to induce abortion. Thankfully, I had eaten a few leaves at most. I began to wonder whether other foods that I had identified as making me sick were also potentially toxic to the pregnancy. That *is not a question that I can answer, but I can explain how my brain knew that lovage was the culprit.*

There are several processes involved in the Cambridge lovage story, although I did not experience them as separate. There was the initial sensory recording of the smell and the taste of the lovage, followed by the formation of a memory of this sensation – otherwise I would not have recognized it again – and then the re-activation of the memory on re-smelling it some days later that brought with it, incredibly, a sense of the sickness that it had caused. All this had happened before the visual identification of the lovage. The sequence of experiences demonstrates that the nausea/disgust feeling had happened before I saw the lovage. It was as if the olfactory memory of the lovage was paused, ready in waiting to be ignited if smelled or tasted again, to warn me of its poisonous consequences. The remembered smell had made a feeling. I was not even aware that I knew the smell of lovage. How clever are our brains?

A smell may cause a range of feelings. Those of the 'vanished loveliness'* [i] of childhood; of love – a baby's scalp; of sexual excitement – a lover's neck; of fear – rancid sweat; of disgust – rotting fish or, in my case, lovage. Smell brings you back in time and warns you about the future in the blink of an eye. How does smell do this in such an immediate way? We need to start at the point of entry of smell sensation into the brain – the odour receptors in the upper nasal passages that recognise different scent chemicals.[29] [ii] The scent chemicals may be in food that is tasted – a madeleine, a herb – or may be airborne – the scent of cut grass, a lupin. At a molecular level, the specific pairing of an odour molecule to its matched odour receptor in the nasal passages triggers an electrical signal that is carried to the brain in a very short nerve, about 5 cm or 2 inches long, called the olfactory nerve. The olfactory nerve runs horizontally from the back of the nose to a structure in the brain called the *amygdala* – the heart of the matter of memory (see Figure 6). I call the amygdala the 'emotional sparkplug' of the brain, because it triggers emotional responses and feelings. It

* William Styron, *Lie Down in Darkness* (Vintage Books, 2000), pp. 51–2.

is situated directly in front of the hippocampus, with which it is densely interconnected, and into which the amygdala weaves its emotional synapses. When neurons link up, as we know, they form cell assemblies that all then fire together. Amygdala–hippocampal connections form the basis for emotional memory.

The Amygdala

The emotional sparkplug, the amygdala, like the hippocampus, is plastic and can easily grow synaptic connections. Also like the hippocampus, it has direct connections with the sensory cortices, in particular the visual cortex, facilitating emotional response to imagery. The difference between smell and the other four sensations is that smell neurons from the nose go *first* to the amygdala – hurtling into it from the short run from the nasal passages – before reaching the olfactory cortex. This is why we experience an immediacy of emotion when we smell. Sensory experiences other than smell – seeing, hearing, tasting and touching – relay through their cortex on the surface of the brain before diving down to the amygdala/hippocampus. Something is seen before you feel the associated memory; you hear a song before you remember the summer when it was a hit. Smell neurons, by first relaying to the amygdala, trigger feelings before the smell can be consciously identified. Smell, as Proust described, is *remembered as a feeling.* It is quite amazing that, through intense introspection, Proust could pinpoint this subjective phenomenological experience before it could be explained by science.

In my case, the lovage molecules had triggered a signal in the nasal passages that had travelled to the amygdala causing a remembrance of a feeling of nausea. Meanwhile, the image of the sunlit lovage was shimmering on my visual cortex. The remembered nausea, the identification of the smell, the visual identification of the lovage and the memory of the salad came together, and the knowledge that the lovage had made me sick emerged from the neuronal mix.

Figure 6. The journey of smell from the nose to the amygdala

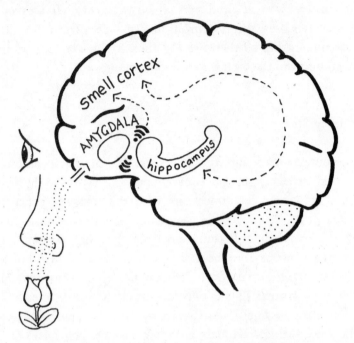

The olfactory nerve carries the electrochemical signal from receptors in the upper roof of the nose through two routes in the brain. (1) the quick run to the amygdala, where the emotion associated with the smell is released, and (2) the longer run to the olfactory cortex, where the smell is identified. The smell and taste cortices overlap and this is why taste and smell are often difficult to distinguish from each other, or, rather, why smell is an intrinsic part of taste.

The Amygdala and Emotions

How can a small brain structure, the amygdala, create the experience of an emotion? We were taught in medical school that the amygdala is the 'emotion centre' of the brain, but for me this seemed improbable and the information had no fit in my personal cognitive constructs. Since then I have built up a memory framework for understanding the emotional systems in the human body, and I now know that the amygdala does not create the emotion but is a nerve centre from which neurons emerge to make emotions in the body. Before looking at how emotions are made in the body, let's look at how we know that the amygdala is the emotion maker.

The most commonly studied emotion in animals is fear. Fear is a well-tried and tested method for measuring emotion, because when animals are frightened they respond in ways that can be visualized and measured, such as running away or freezing. *Emotions* cause *motion*, and motion can be measured. The most famous amygdala studies were conducted in the 1930s and 1940s by two scientists, Heinrich Klüver and Paul Bucy. Their names are well known to students of brain systems because of the Klüver–Bucy syndrome, caused by removing both the right and left amygdalae in monkeys. Bucy, a neurosurgeon, removed the hippocampi and amygdalae from both hemispheres of the male monkey brain (this was before the days of animal rights). Klüver, an experimental psychologist, observed that, following the surgery, Monkey lost fear behaviours. Because Monkey no longer, presumably, experienced fear, he no longer behaved in an appropriately passive and submissive way to the dominant, and therefore stronger, male in the colony. Nor did he learn fear from his blunders into inevitable defeat. This led to serious injury, social isolation, and eventual death. In a world without fear, Monkey perished.

If you were a monkey who had been spared a bilateral loss of your amygdalae from Bucy's surgical knife, and still had nice healthy amygdala, and if you were confronted by the dominant monkey, your heart would beat fast and hard, your pupils would dilate, your muscles tighten, your breathing quicken, your blood pressure rise, and the stress hormone cortisol would be secreted. These physiological responses constitute the emotion of fear. The emotions happen because the amygdala is alive and responsive, unlike those of your unfortunate colony member whose amygdalae have been lesioned, and who knows no fear. Female monkeys were less studied, but the reports interestingly indicate that those who had no amygdalae had impaired maternal behaviours, frequently abusing or neglecting their offspring. The experiments showed that fear, including maternal anxiety as I speculate, was mediated through the amygdala in monkeys and was necessary not only for individual survival, but also for the survival of the colony.

Humans can, although very rarely, be afflicted by the disease called Urbach–Wiethe, in which the amygdalae become damaged, leaving the surrounding brain unaffected. This leads to a failure to recognize fearful facial expressions, and a reduced ability to register fear overall.[30] Non-functioning amygdalae means that the person can make event memory, but it will not have a normal emotional content, and will not be recalled with the associated emotion.[31] On the other hand, those with intact amygdalae but damaged hippocampi can feel the fear but are unable to consistently form event memory and avoid the stimulus that triggered the fear. The consequences of not having a functioning amygdala are sometimes dramatically observed in Urbach–Wiethe sufferers.[32] The life of one person with this disease, known as 'SM', has been documented in the academic literature.[33] SM has been reported as having normal event memory, but she does not register feelings of fear, even in life-threatening situations, and she does not learn to avoid harm from previous experience. She approaches strangers without any sense of fear and tends to stand in very close proximity to them. Because of her inability to sense fear cues, and her behavioural disinhibitions, she has had several near-death experiences that have not led to future avoidance of these dangerous situations. She seems unable to feel fear and to learn from fear. Interestingly, she describes feelings of curiosity when an average person would be feeling fear – for example, she was curious about what a tarantula would feel like to touch.

The amygdala lights up in MRI neuroimaging when a person is feeling fear, such as when challenged with a hypothetically threatening scenario.[34] Individuals with spider phobias have very strong amygdalar activation when exposed to pictures of spiders compared to those who are not arachnophobic.[35] If we were able to hypothetically peer into the brain of an individual with arachnophobia as they were looking at a picture of a tarantula, we would see connections lighting up between the location on the visual cortex where the tarantula is mapped and the amygdala.[34] This hypothetical pathway is both memory and present experience. It is made from past experience and is creating new emotion. The next big question is, how does the amygdala generate emotion, in this case fear?

Figure 7. The amygdala as the emotional sparkplug

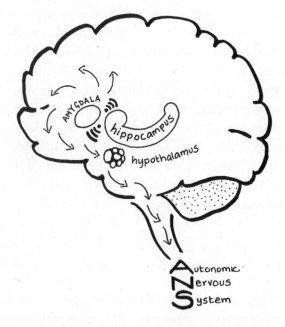

This illustration demonstrates how the amygdala circuits to the hypothalamus to ignite the ANS in the body and create visceral emotions.

In essence, neural outputs from the amygdala go to the body and make feelings in the body. The theory that emotion arises in the body was first proposed by a hero of mine you may remember from Chapter 1, William James, in 'The physical basis of emotion' (1894).[36] [iii] William, his more famous brother Henry, and his less famous sister Alice, were all magnificent interpreters of human emotions: James as the master of fiction, William as a psychologist, and Alice as a diarist who wrote vividly about her emotional breakdowns and depressions. William James proposed that feelings were caused by the activation of visceral sensations in the body. We now know that James was correct and that the amygdala is directing this visceral activity from the brain, hence my name for the amygdala, the emotional sparkplug. The system sparked in making emotions is the autonomic nervous system (ANS), which innervates all inner body organs: the heart, the gut, the lungs and blood vessels, as well as the skin and some glands and smaller muscles. Functions mediated by the ANS include facial blushing and blanching, dilation and constriction of the pupil, respiratory rate, heart rate, tear production and sexual arousal. 'Autonomic' is a synonym for 'automatic': our ANS function is generally thought of as, and generally is, automatic and mostly out of our control. The heart doesn't beat, or the gut contract, or blood vessels dilate, because we will them to do so – these things are automatic. Nevertheless, it is possible to learn to modify autonomic functions through meditative work, and this is the basis for mindfulness.[iv] The ANS is like a puppet on strings, a marionette, that is tugged this way and that by outputs from the brain.

The Hypothalamus

The more precise marionettist of the ANS is the *hypothalamus*, a small collection of densely packed neurons, with the right cluster only separated from the left by a small canal of brain fluid that marks the centre of the brain at the level of the bridge of the nose. The hypothalamus is very close to the amygdala, to which it is hard-wired.

Anatomically, multiple brain circuits, one of the most important of which is the amygdala/hippocampus, converge on the hypothalamus, and it is the sum of these inputs that will determine the output to the ANS. Darren Roddy, a psychiatrist and a neuroimager who works in our group, says that he sees the hypothalamus as the place where all outputs from the memory-emotional brain converge and then go to the body to bring about changes in the ANS and the endocrine systems. Multiple interoceptive body systems are modulated through the hypothalamus – it being the control centre not only for the ANS, which makes feelings and emotions, but also for output from the brain controlling the cortisol stress system in the body. I have spent most of my research career looking at the cortisol stress system, pondering how many ways the brain can zap the hypothalamus to change the way that we feel. The hypothalamus is the final exit point to the interoceptive body, and the flow is not just from the brain to the body, changing the interoceptive body: it also occurs in the other direction, with the interoceptive body changing the brain. Cortisol and stress will be the topic of a future chapter, but for now just note that emotions and stress have the same brain marionettist, the hypothalamus.

In the pathway from the sparkplug of the amygdala, we have now passed through the exit door of the brain, the hypothalamus, and are travelling through the ANS to the interoceptive body.

A Rainbow of Feeling States

The strength of an emotion can be measured by looking at ANS arousal. Visual stimuli are more likely to provoke a bigger ANS response than auditory stimuli. This can be seen in brain anatomy, where one sees bigger input from the visual cortex to the amygdala compared to the other sensory cortices. This is fascinating when you consider that the original Sensationalists of the eighteenth century, including Locke and Molyneux, were preoccupied with sight and visual knowledge/memory rather than other sensations. Perhaps Molyneux and Locke and the other pre-Enlightenment philosophers intuited the highly interconnected visual-emotional circuitry when they selected sight to demonstrate connections between sensation and memory. Combined visual and auditory stimuli will provoke a larger ANS response than either alone.

Most of us feel uncomfortable telling a lie – this uneasy feeling is generated by ANS arousal.[v] The stereotypical lie-detection needle that oscillates in movies telling us that a person is telling a lie is measuring ANS arousal, and specifically whether a person is sweating, because it is a reliable measure of arousal. The ANS emotional system can create a range of opposite feeling states: a quickened or slowed heart rate – arousal or relaxation; an increase or a decrease in blood pressure – tension or swooning; dilation or constriction of the small blood vessels of the skin – flushing or blanching; gut immobility or overactivity – bloating or rumbling. The range of opposing emotional states can happen because there are two ANS body systems in the body – the *sympathetic* and the *parasympathetic* systems, both controlled from the ANS HQ in the hypothalamus. As a general rule, activation of the sympathetic system causes increased activity in the innervated tissues and organs, such as heart palpitations, muscle tightening, sweating, fast breathing or increased blood pressure. This is often called the 'fight or flight' system. Activation of the parasympathetic system, on the other hand, slows the heart, reduces blood pressure, reduces gut movement and reduces blood flow to the skin, and it is commonly referred to as the 'rest and digest' system.

There can be a range of activation of either the parasympathetic or sympathetic systems allowing for a range of body system arousal and emotions. Many of the most intense human emotions are mixed. Let us look at one of the most powerful and well-documented emotional experiences in the large repertoire of documented human passions. Alain-René Lesage's description in *L'Histoire de Gil Blas de Santillane*, written between 1715 and 1735, of the beginning of Seraphine's love affair with Don Alfonso is a romantic classic.

It was quite dark and the rain was pelting down; I had crossed several alleyways and suddenly came upon the open door of a drawing room. I went in and at once became aware of the magnificence of the palace . . . on one side I saw a door a little ajar. I half opened it and could see a vista of rooms, the last of which was lighted . . . Then I noticed a bed, whose curtains were partly drawn apart because of the heat, and my attention was riveted by the sight of a young woman who lay asleep . . . I moved a little closer . . . I felt overpowered . . . While I was standing there, dizzy with the pleasure of looking at her, she awoke.*

This description of the heady mixture of arousal and pleasure of such a *coup de foudre* could have been written yesterday. Emotions are constant across cultures and time, indicating that the biological machinery for feeling states is universal. When Don Alfonso saw Seraphine *his attention was riveted*, he felt *overpowered* by emotion, *dizzy* with *pleasure* . . . the intense sympathetic and parasympathetic synchrony of love-at-first-sight.

In 1812, Stendhal, a gifted and seemingly dispassionate observer of the passions, wrote a wonderful book, *Love*, in which he quoted this passage from *Gil Blas* as an example of the 'birth of love'. The following wise words from his book say so much about the experience of romantic love: 'Nothing is so interesting as passion; everything about it is so unexpected; and its agent is also its victim.'†
As Stendhal observed, we may be the happy victims of a *coup de foudre*, or the unhappy victims of unrequited love. We are victims because we do not will it – it happens. How can an overwhelming emotion like a *coup de foudre* just happen? To understand at least some of the answer to this question, we need to invoke memory. Back in the seventeenth century, Descartes came to understand personally the effects of memory on romantic feelings when he learned, through self-observation, that he was attracted to women with cross-eyes. He worked out that he had fallen in love with a cross-eyed girl when he was a boy and that the image of cross-eyed women

* https://www.exclassics.com/gilblas/gilblas.pdf (p. 199).
† Stendhal, *Love*, trans. Gilbert and Suzanne Sale (Penguin Books, 1975), p. 219.

was triggering the emotional response. He was recognizing in this our tendency to be guided unknowingly by emotional memory. Many of us, after all, do marry our 'fathers' or 'mothers'. It is memory that partly makes us the unknowing victims of our passions.

But still, a *coup de foudre*, although very compelling, is a relatively simple feeling in the sense that it is an immediate and instantly recognizable blast of mixed interoceptive sensations. The range of descriptions of 'heartfelt' emotions is wide and sometimes vague – your heart may feel heavy, or light, it may be bursting with happiness, it may be broken, it may pound, it may feel as if it has stopped for a moment, or something may just tug on it that you cannot immediately identify. At other times we feel frankly 'mixed up' and complicated and are overcome with feelings that we are unable to interpret. Our body may be telling us something, but what is it? William James knew that human feelings were more than body sensations when he defined the experience of an emotion as being an *interpretation* of a physiological sensation that arises in the body. An emotion is 'not a primary feeling, directly aroused by the existing object or thought, but a secondary feeling indirectly aroused'.* To paraphrase, a primary feeling is a physical sensation in the body (the ANS), and a secondary feeling is the interpretation of this as a calibrated emotion of fear, love, disgust, and so on. For example, when you are about to go to a very important job interview, and you feel your heart racing and butterflies in your stomach, you know that you are nervous or excited and that you are not falling in love. The heart racing and the butterflies are the primary feeling, and the understanding of this as being fear and anxiety in the context of the impending interview is James's secondary feeling. The secondary feeling is the interpretation of phsyiological changes that have occured involuntarily in anticipation of the interview.

Now we are at the real heart of the mysteries of the passions – it

* William James, *Principles of Psychology* (1890; reprinted Dover Publications, 2014).

is not just the making of the feeling that is important, it is how we interpret it, or sometimes fail to.

As alluded to in Chapter 2, all of the interoceptive sensation coming from the inner body – the heart, gut, lungs, sexual organs, blood vessels – are mapped in a small piece of cortex hidden under the surface of the brain called the insula. We need the body – the ANS – to make the feeling and the insula to interpret the feeling.[37]

The Hidden Cortex, the Insula: The Rag & Bone Shop

The insula is so named from the Latin for island (plural, *insulae*), because it resembles an island of cortex inverted into the brain. Try to locate your own insula anatomically: presumably you are holding this book as you read it, so, with your free hand, put the tips of your fingers on the side of your head from a point starting at the upper junction of your ear and scalp and with your other fingers at a slight angle upwards and backwards. I want you to imagine that you can push the brain tissue lying under your fingers into your brain, as if you were pushing in on a deflated football, creating an indentation in the surface of the brain cortex. The squashed-in part of the cortex is the insula (see Figure 8).

One way of examining insular function is to look at individuals who have disorders in the insula. The study that I'll cite, following on from my lovage story, is about the emotion of disgust, and it hails from California. The authors reported on the experiences of nausea, or disgust, in patients with degenerative disease of the insula.[38] Insular atrophy occurs commonly, for example in Alzheimer's dementia. They reported that the loss of disgust sensation in the patient with Alzheimer's was related to reductions in insula volume on neuroimaging measures. This seemed to be quite specific, because there was a proportional relationship between the ability to experience feelings of disgust and insular sizes – the smaller the insula, the less able were the subjects to experience disgust. People with a diagnosis of anorexia nervosa tend to have a poor

Figure 8. The insula

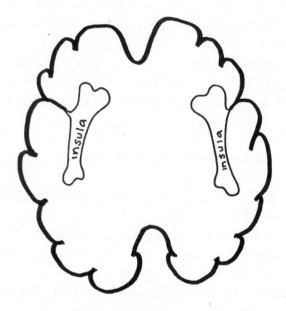

A cross-sectional cut through the brain at approximately the level of the top of the ears, showing the hidden cortex of the insula.

understanding of their internal states – this can be seen in their inability to appreciate the sensation of hunger or of feeling satiety. Although no abnormality is present in the volume of the brain insula in anorexia, there is reduced activity in this region in response to changes in internal feeling states.[39] Conversely, those suffering from depression, who experience overwhelming and negative emotional states, have been found to experience more activation of the insula in response to facial expressions of disgust.[40]

Wilder Penfield published a paper in 1955 in which he described the effects of stimulating the insula in conscious patients with exposed brains who were about to undergo surgery for epilepsy.[41] He noted that patients experienced a gut sensation during stimulation of the insula. Penfield's studies became a live clinical issue for me some years ago when treating a patient whom I'll call Stella.

Stella

Stella was referred to me by her GP following investigations by multiple medical specialists over many years for investigation of a 'queer' sensation in her abdomen. She had consultations and opinions from multiple disciplines, including general and gastrointestinal medicine, neurology and gynecology. They had all reported that they could find no abnormality with Stella's gut, either in terms of structure or the mechanics of gut motility – there was no tumor leaning on her gut, no peripheral neuropathy (nerve disease) or gynecological abnormality. She complained of a very specific and unpleasant sensation – a feeling of electricity in her abdomen extending to her chest. She called this 'the electrical buzzing'.

At our first meeting she told me that the buzzing had been present for many years but was getting worse. She was now at a point where she would try anything, including coming to see a psychiatrist, to get rid of it. She couldn't understand how a psychiatrist could help, but she was following her GP's advice. The electrical buzzing had been present for so many years that she had begun to believe that someone might be doing it to her, that it was perhaps being planted in her body by someone or something outside of herself. Whatever I asked Stella about her buzzing sensation, she would return to her husband's smoking habits: she hated her husband smoking in the house. This seemed reasonable, as Stella was not a smoker, but what had her husband's smoking got to do with her buzzing gut sensation? She was particularly angry because of the ashtrays, she told me. I wondered if he was not bothering to empty them. She began to elaborate on how he left the ashtrays around the house and the arrangement of the cigarette butts that they contained. I persisted in trying to understand. Eventually it transpired, as I interpreted it, that she thought that her husband was transmitting messages through the cigarette butts and that the pattern of the discarded butts in the ashtrays represented some sort of coded communication.

The cigarette code was being shared by others, particularly friends of her husband who visited the house to drink whiskey and play cards, and who also smoked. They were conspiring to harm her, but she did not know why or how they were doing this. There was other evidence: the position of the furniture sometimes changed, a magazine that she had left open would be closed when next she saw it, the milk would be on a different shelf in the fridge, and so on. She appeared to be misinterpreting and over-interpreting events. Stella's husband told us that Stella had been talking about ashtrays in mystifying ways for about fifteen years.

Stella had a long-standing psychotic disorder that was finally diagnosed because of the electrical buzzing sensation in her abdomen. We call this a somatic hallucination – a sensation experienced as coming from the body that is probably arising in the brain. A somatic hallucination, although infrequent in psychotic states, is one of the key experiences that indicate a diagnosis of schizophrenia. Sensation from both the exteroceptive, the Big 5, *and* the interoceptive, the visceral, organs can be generated from within the brain.

I explained to Stella as best I could that the electrical buzzing was not being caused by something in her abdomen but that it was probably being caused by mis-firing or mis-wiring in her brain. I also told her that we may be able to control the buzzing with antipsychotic medication. Stella thought that this was all a bit strange, but she was willing to try anything and commenced a trial of antipsychotic. The buzzing faded over the following weeks and left her completely some months later. She gradually stopped believing that her husband was in cahoots with his friends to try and harm her, or that they were leaving signals for her in the ashtrays, or that they were rearranging objects to let her know that they were monitoring her. She wasn't interested in the ashtrays any more, and if I drew her attention to her old preoccupations she shrugged it off. It had happened, now it did not. Stella felt no need to retrospectively reconfigure her memory in the light of new events. She just wanted to take the medication and remain in her familiar, uncomplicated world without torment from the distressing abdominal sensation. She began to come back to her old self that had seemed long gone, communicating again with her family and neighbours, doing domestic chores and tending to her neglected garden.

I could find no reference for Stella's peculiar somatic hallucination in the psychiatric literature. Some time later I read a 2009 paper written by a neurosurgical group led by D. K. Nguyen from the Notre Dame and Saint Justine hospitals in Montreal.[42] They reported on sensations experienced by conscious patients, prior to neurosurgery, following stimulation of the insula. They noted in one of their papers a peculiar sensation that a patient described as a 'buzzing in the abdomen' produced by stimulation of a particular patch of the insula. On further investigation I found that Penfield's seminal 1955 paper had described a similar singular 'buzzing' sensation in some of his preoperative patients on stimulation of the insula. I was struck immediately by the similarity of Penfield's and Nguyen's descriptions, decades apart, to Stella's singular visceral sensation. The neurosurgical experiments show how stimulation of the insula from *within* the brain can cause a buzzing gut feeling. That the same part of the insular cortex activated manually and deliberately by Penfield and Nguyen was being activated in Stella, by some sort of cortical pathology, seemed, and seems, likely to me.

So, in practice the insula is a sensory cortex of inner visceral sensation that lights up when we experience emotion. A study published in 2004 by Ray Dolan and colleagues in London demonstrated how, in normal emotional states, the insula lights up when the heartbeat is being subjectively monitored.[43] They looked at brain activity in subjects who were also questioned about their awareness of their heartbeat and found that an individual's level of interoceptive awareness was commensurate with the size and with the activity of the insula.[44] In the late 1980s Antonio Damasio proposed that even complex human feeling states could be 'mapped' on the insula. Damasio theorized that internal body sensations could be organized into countless combinations to give 'a range of feeling states'. In his book *Self Comes to Mind**
Damasio has written lyrically about his neuroimaging experiments in which his research group demonstrated that different emotional states were associated with different areas of insular activation, which he called 'emotion-specific neural patterns'. The left insula was predominantly activated in positive emotions, such as maternal and romantic love, listening to nice music or happy voices, smiling or watching

* Antonio Damasio, *Self Comes to Mind* (Vintage, 2010).

Figure 9. The insula as the emotional cortex

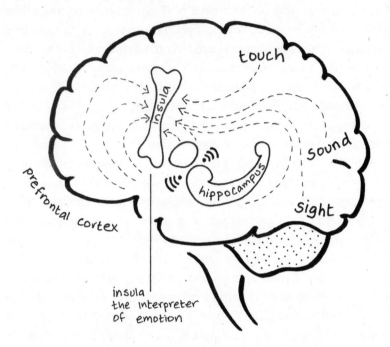

Some of the main pathways that converge on the insula from inside the brain. The insula is a sensory cortex that records emotion states from inside the body but can also be stimulated from inside the brain, allowing memory pathways to cause emotional states.

others smile; even the positive emotional states associated with feelings of anticipated purchases, possibly the neural basis of 'retail therapy', are related to activation of the left insula.[45] The emotional mapping explains the range of feeling states that can be experienced by humans, from the *coup de foudre* of Don Alfonso to the mysterious ghost of an emotion stimulated by Proust's madeleine.

Life's Memories and the Insula

There are pathways connecting different areas of the brain to the insula. If a part of the insula is fired off by a neuron from another brain area, this will cause the experience of an emotion. These brain pathways, forged from memory, cause feelings, much like Penfield or Nguyen stimulated the insula with surgical instruments to produce feeling. The neural paths from the prefrontal biographical networks to the insula means that biographical memory inputs can stimulate feeling states (see Figure 9). In this case, where emotion is being stimulated from biographical memory, the amygdala does not have to be involved: the insula is being stimulated from inside the brain by memory neurons from the prefrontal cortex.[46]

The more we learn about the insula, the more we understand not just how emotions and past biographical memory are neutrally woven, but also how society impacts on the emotional wellbeing of the individual. I was particularly drawn to one study that looked at the brain activity recorded in those who felt socially excluded, like many patients who suffer from psychiatric disorders. The authors found that the representation of the pain of social exclusion was mapped adjacent to that for physical pain. Social exclusion 'hurts' . . . who says that there is no such thing as society, or that politics is not personal?[47]

Life inevitably involves loss and difficulties, and emotions during crises seem to be all amygdala – intense, overwhelming, certainly automatic and perhaps even uncontrollable, like the *coup d'état* of psychosis or the *coup de foudre* of Don Alfonso. As biographical memory goes cortical over time, so the emotions seem to modify, perhaps reflecting the shift from a highly aroused amygdala drive to the more considered prefrontal-insula drive. This is speculation on my part, but it would explain immediate and more measured feelings. The transition from the experience of raw emotional states driven by the amygdala, to a recollected one, more likely to be driven by the prefrontal-insula circuit, is reflected in the universal experience of the death of a loved one. The emotions experienced immediately following a death are intense, sometimes uncontrollably so, distressing and painful. The grief intrudes and dominates – the amygdala is on fire, branding all sensory input with the searing imprint of the loss. Everyone and everything is a reminder to the grieving person of the loss: 'every old man I see reminds me of my father,' as Patrick Kavanagh wrote following his father's death.* The memory over time goes cortical, becoming represented in the prefrontal biographical networks and the softer insular feeling. The memory is separating from the hammerjack of the amygdala as the grief transfers slowly to a prefrontal-insular emotion, one in which the bereaved person is more likely to experience pity for the dead loved one . . .

> A pity beyond all telling
> Is hid in the heart of love.†

Living and learning is a never-ending dance of sensation, memory and emotion. Sensation from the outside world gets woven into cortical maps of sensory experience, and events get woven additionally into the amygdala–insula loom of the emotional circuitry. Ultimately, what is memory without emotion – an endless repertoire of experiences without any human meaning. And emotion without memory? A shallow flitting from one object of desire to the next. Without emotions our hearts would not break, we would not grieve, but nor would we have memories rich with the people that we were drawn to and lived life with, even if only temporarily; the sort of memories evoked by meeting the cousins that you have not seen in years.

Remembering the lovage story now makes me feel something very different from the faint amygdala-driven disgust that I experienced all those years ago. The emotion snagged to this memory is now a diffuse, insular one. It is a hazy coming together of cumulative memories. I remember the balmy heat, a sense of languid anticipation, a 'pre' innocence of what was to come; and suffusing all this was the melancholic nostalgia of remembered cherished moments. The neural magic underlying the sense of vanished loveliness remains as something that *feels* immaterial and soulful.

* Patrick Kavanagh, 'Memory of My Father'.
† W. B. Yeats, 'The Pity of Love'.

6. A Sense of Place

In Chapter 4 we looked at the never-ending neural re-constructions of the plastic dendrites in the hippocampus to make cell assemblies that are then woven during the night into the more stable cortical networks, like Rumpelstiltskin weaving straw to gold while the miller's daughter slept. In Chapter 5 we saw how the amygdala engages in a dendritic dance with the hippocampus, arborizing feelings to hippocampal memories, before coming to the more restful repository of the prefrontal-insula circuitry. Knowing about the neuronal enmeshment of feeling with memory, and keeping this in mind, I want to return to the coordinates of memory construction – time, place and person. Of these coordinates, place has historically held the dominant position. Even this word 'position' – like 'topic' (from Latin for 'place', *topos*), 'commonplace', 'situation' (Latin for 'to place', *situare*) – demonstrates how language has evolved to reflect the central importance of place in past and working memory. 'Working' memory is what neuroscientists call working things out or thinking. A simple example of the primacy of place in memory is the natural inclination to place oneself somewhere at the time of a momentous event. Why do we ask *where were you?* when something happened?

Where were you when you heard about 9/11? At the time of writing this I get over 500 million hits on an internet search of this question. My first memory of where I was when a tumultuous event happened is probably when John F. Kennedy died in 1963. I have a memory snapshot of an event that I often suspected occurred on the day that he died. Luckily, I had never told anyone about this snapshot, so I decided to become the subject of my own experiment, writing down my memory and then asking my mother about the circumstances. The memory was that I was a pre-school child in the back garden facing the escallonia bush that separated our garden from that of our neighbour. I was standing alone by the back corner of the bush, where the hedging was worn thin from our crossing into our neighbour's garden and the wire fencing was visible. My mother was approaching from our house. Our neighbour Mrs Begley was hurrying down her garden path towards my mother, with her hands on the sides of her head in a gesture of distress. The next scene is of the two women talking in an agitated manner, embracing and comforting each other.

I asked my mother if she remembered where she had been when she heard of the death of JFK. She remembered that we were living in Orchardstown Drive in Dublin and hearing it on the 'wireless' in the kitchen and then going to the back garden, where the mothers gathered spontaneously to share the shock. My mother couldn't remember whether I was there or not, but she did confirm that I would have been at home because I only started school in 1964. The memory was just a moment, nothing more than an observed moment with a queer emotion, to which no event was attached. I can now identify the emotion I felt back then as being that of a witness to something from the adult world that I did not understand and from which I was excluded.

I can still remember the interior of that house in Orchardstown Drive that we left when I was six years old, fifty-five years ago. Most of us do remember our childhood homes, in which we house our most distant memories. Going back to one's childhood home is like exploring childhood memory, what Gaston Bachelard called 'psychogeography'. Bachelard (1884–1962) was a French philosopher and architect whose work explored what he termed 'the intimate space' of the domestic home. In his best known book, *The Poetics of Space,* he guides the reader through the central idea that humans create intimate places through memory, often the family home of origin, where they feel emotionally safe and free to create and imagine. His work points to the centrality of place in memory, and thus in imagination: what he called 'the poetic'. Memory, as I will discuss in the next section of the book, is the substrate of imagination.

House-of-origin memories are not always safe, though, and can provide a metaphor for disturbed memories – in our dreams, and in literature. 'Last night I dreamt I went to Manderley again' is the famous opening line of Daphne du Maurier's *Rebecca*. Rebecca's memory is literally 'housed' in Manderley – perhaps that is why it had to be destroyed before the new Mrs de Winter could establish herself. The metaphor of the house burning to the ground as a liberation from torturous memories was borrowed by du Maurier from *Jane Eyre*, written by Charlotte Brontë a century previously. The central metaphor in this book *is* Thornton Hall, where the memories of Rochester's first marriage are played out in the hauntings of the imprisoned 'mad-bad' woman in the attic. Thornton Hall is burnt to the ground, liberating Jane and Rochester from these memories, freeing them to pursue their own dreams. The motif of the haunted house crosses times and cultures and is, like all myths, fertile ground for psychosis. Anita's story, below, of her haunted home sparked many associations in my deep and recent memory, resonating with myth and the cruel reality of the lives of working women in pre-feminist times.

Anita

Anita was in her seventies and attended our day hospital five days a week for years. She had a long-term psychotic illness that seemed to have commenced following the birth of one of her children. At this point she was hospitalized, and the course of her psychosis waxed and waned for the next twenty to thirty years. She nevertheless struggled through mothering her children, almost as a lone parent as was the way in the 1950s for most wives. Her husband had a solid job and 'brought home a good wage'. No more was expected of husbands and he enjoyed pints with his mates after work a few nights a week, and had a family day on Sundays, with roast lunch and sometimes a GAA (Gaelic Athletic Association) match in Croagh Park. *The GAA is a hugely influential national association, built around the native Irish games gaelic football and hurling, that engenders loyalties to local county clubs that are as passionate as those in English football or American baseball, except that in Ireland every parish is involved, and it is not professional.* Anita's personal history was like a social history of Ireland, with the invisible wife/mother that no one seemed to notice until she was no longer able to manage her domestic work. I imagine her, face turned to the wall, washing dishes after the Sunday lunch, seemingly quiet and content to be serving her family, while she was really suffering, in silence, from the unending distress of a psychosis. She was a heroic survivor not only of the casual neglect of the women of her day, and of her psychosis, but, it transpired when we later treated her, also of childhood sexual abuse.

Her enduring delusion was that her house was haunted. She could hear whispers at night, and in the morning she could see that the ghosts had moved something slightly or switched on a light downstairs that she remembered switching off before going to bed. Once or twice she had felt something touching her in bed. She was being monitored in the house, being watched and listened to, and she knew that she had to be careful about her actions. If she was not at home she did not have these experiences. She seemed fine in the day hospital, very reserved but not distressed, and one would get the impression of a woman with a gentle mien dipping in and out of a benign brain disorder . . . until you talked to her and discovered the haunting psychosis. In an effort to remove her from the trauma of the family home, her daughter brought her to live with her family. Over a short period of time she came to believe that something was amiss in her daughter's home, and became delusional about strange happenings. Anita then returned to live in her own home and refused to move again until her sudden death at home from a catastrophic stroke. The day hospital shut down for a day to attend her funeral.

Places, places, places . . . Why is place so important in memory? Perhaps place memory is an evolutionary legacy from times when remembering a location – for example where food could be foraged successfully or where danger lurked – was critical to survival. The

great French sociologist Maurice Halbwachs, in 1950, put his observations of the central importance of place in recall in his landmark
book *The Collective Memory.**

> Now let us close our eyes and, turning within ourselves, go back
> along the course of time to the furthest point at which our thought
> still holds clear remembrances of scenes and people. Never do we go
> outside space. We find ourselves not within an indeterminate space
> but rather in areas we know or might very easily localize, since they
> still belong to our present material milieu. I have made great efforts
> to erase that spatial context, in order to hold alone to the feelings I
> then experienced and the thought I then entertained. Feelings and
> reflections, like all other events, have to be resituated in some place
> where I have resided or passed by and which is still in existence. Let
> us endeavour to go back further. When we reach that period when
> we are unable to represent places to ourselves, even in a confused
> manner, we have arrived at the regions of our past inaccessible to
> memory.

Considering this centrality of place in memory it is not surprising that the most important cells in the hippocampus are the cells
that recognize place and are appropriately called 'place cells'.[48]

Place Cells

I travelled to Cork by train in August 2015 to hear the neuroscientist
John O'Keefe give a lecture about his discovery of place cells in the
hippocampus. O'Keefe, along with May-Britt Moser and Edvard
Moser, won the Nobel Prize in Physiology or Medicine in 2014 for
their discovery of place cells.[49] John O'Keefe sounds American,
lives and works in Britain and is a young-looking older man, wiry

---

* Maurice Halbwachs, *On Collective Memory*, ed. and trans. Lewis A. Coser (University of Chicago Press, 1992).

in build, with pale blue Irish eyes that darted around as he delivered a lively lecture. His talk was to become another dawning for me. I remember later that day standing at a bus stop outside University College Cork with a sense of a cognitive shift. I had known about place cells, but they only came alive for me in an experiential way when I listened to O'Keefe's journey of discovery.

His story of the rat hippocampus and place cells started with his experiments using live rats moving around an enclosed space, the experimental field. Rat had had a micro-wire inserted into a single neuron in the hippocampus – an amazing feat in itself, considering that there are about 180,000 hippocampal neurons in rats – and Rat was alive and moving freely. The micro-wire was connected to a machine that recorded the electrical activity in the punctured hippocampal neuron. This allowed O'Keefe and his team to see what was happening in a single memory cell in the buried hippocampus of Rat. What would happen in this lone neuron as Rat moved around the experimental field? The field was broken down spatially into small squares, like a map grid. When Rat moved into a specific square a signal appeared on the recording machine. This happened again when Rat re-entered this square, but no other square. The neuron 'recognized' that specific square, and no other one. O'Keefe called these neurons 'place cells'. A place cell was the unique identifier, or cell memory, of a specific external place: one hippocampal neuron equalled one grid square. Remember, this is not the cortex, the visual cortex or otherwise, it is the hippocampus – the memory machine.

This stark demonstration in 1971 of a connection between a small and exact place and a single memory cell in the hippocampus had profound implications.[48] The function of the hippocampus as a specialized memory centre, distinct from sensory functions, was now established. Rat was seeing, hearing, touching and smelling, had full body sensory function mediated through the outer cortex. All of these sensory functions were taking place in the cortex when Rat moved around the experimental field – seeing it, smelling it and touching the experimental field floor – while the hippocampal neurons were making place memory through a meticulous process of place-in-the-external-world-to-place-cell. Another finding that emerged from later work using these techniques was that places that were adjacent in the external world did not fire adjacent place cells in the hippocampus. It became apparent that different patterns of firing place cells represented different locations, but the patterns of firing neurons were abstract, not geographical, representations of the external spatial structures. Places seemed to be represented by the codes of cell assemblies.

John O'Keefe's type of intracellular recording of a hippocampal cell cannot be done in the human hippocampus under normal conditions, but intracellular recordings have been done opportunistically in hippocampal cells in conscious patients prior to them undergoing surgery for epilepsy. A clinical study published in *Nature* in 2003 examined electrical responses from *individual cells* in seven such patients.[50] Participants underwent a test in which they were first familiarized with a virtual town in a computer game. The second part of the experiment involved the patients having to navigate a taxi through the memorized town to a specific location. Certain cells in the hippocampus responded repeatedly to specific locations, demonstrating place selectivity. It is comforting to think that we have cells buried in our brains that are steadfastly rooted in place, anchoring us while the world whirls around.

Studies of the hippocampus implicate the right hippocampus in place memory. We looked earlier at the study of the London cab drivers, who have to memorize extensive maps of the big city before being given a cab licence. Neuroimaging demonstrated that it was predominantly the right hippocampus that grew with expanding geographical knowledge.[18] What is even more intriguing is the finding that, when experienced London taxi drivers *imagine* a complex route around the city, the right hippocampus selectively fires.[51] Here we can see that remembering and imagining involve the same circuits in the brain, and we will expand on this in later chapters.

Let us briefly return to MM, my patient who had lost her

memory through destruction of her hippocampi and who, although able to navigate a room when in it, once she left this room could not find her way back. Her hippocampal memory for place was gone, but her sensory abilities in terms of seeing the room and so on were intact. Without the cortical and the place-memory systems working *together*, one is lost. There has to be a system of sensory information feeding into hippocampal memory for full perception to occur.

The seminal experiments looking at how sensation feeds into the hippocampus to make place-cell memory were conducted by Edvard and May-Britt Moser, the wife and husband neuroscience team who shared the Nobel Prize with John O'Keefe. In 2005 they published their discovery of cells in the entorhinal cortex, where integrated sensory information from the different cortices is delivered to the hippocampus. They discovered that neurons from the entorhinal cortex zap the hippocampal cells with high-frequency currents, firing the hippocampal cells to make the dendritic protein that wire the cells together to form the cell assemblies, the place codes.[52] The entorhinal cells feed the cortical threads of sensation into the hippocampal loom.[53] And, during the night, Rumpelstiltskin-like, the hippocampal cell assemblies weave their magic into the fabric of cortical memory. The place cells, over the years since their discovery, have now evolved into 'space' cells, because objects are seen in a spatial three-dimensional context, rather than just a flat two-dimensional place location.[54] We will come to this in the next chapter – just note that from now on I will use 'place' and 'space' synonymously.

What is chosen to be zapped from the cortical dance of sensation into the hippocampus, or what is chosen to be zapped back to the sensory cortices, for more permament cortical storage, or to the swelling stores of the prefrontal cortex as your life is lived, is an infinity of possibilities. Similar to life experience, it is a messy entanglement of sensory cortical arousal and hippocampal memory codes enmeshed in amygdalar tangles. It is the interconnectivity of the whole working in unison that creates experience and memory.

*How can we know the dancer from the dance?**

The Amygdalar Tangles

Whether we are following the psychogeography of our home of origin through our memories, or the current home that we don't want to leave because it houses our memories, place is navigated primarily by images and sight. The visual cortex, as we have explored, is more connected to the amygdala than other sensory brain areas, and when we negotiate place in biographical memory we are also stimulating emotional memory. The tethering of emotional memory to place is beautifully laid out in the book *Suspended Sentences*, by Patrick Modiano, who won the Nobel Prize in Literature in 2014.

At times, it seems, our memories are like polaroids. In nearly thirty years, I hardly ever thought about Jansen. We'd know each other over a very short period of time. He left France in June 1964 and I'm writing this in April 1992. I never received word from him and I don't know if he's dead or alive. The memory of him had remained dormant, but now it has suddenly come flooding back this early spring of 1992. Is it because I came across a picture of my girlfriend and me, on the back of which a blue stamp says 'Photo by Jansen. All rights reserved'?

The beautiful novella, one of a trilogy that makes up *Suspended Sentences*, from which this quote is taken, is a journey through memory triggered by an old photo: everyone is elusive, relationships are fluid, but the named streets and cafés in Paris where the experiences

* W. B. Yeats, 'Among School Children'.

of the summer of 1964 had been lived form the threads through which the reader is taken on the young Jansen's emotional 'memory lane'. It is the summer of the narrator's first loves, with a woman and with photography. The author seems to be living and journeying through that summer in the filtered world of memory and feelings, being navigated by place. Place is fixed in Parisian locations in which the ghostlike encounters are staged. The novella is, fittingly, called *After-image*. I love this story, because it captures that enduring sense of place that is composed of the hauntings of human lives trapped in the glorious and inglorious conditions of their time. The human lives come and go, leaving remembrances – a torn lace curtain visible through a broken window, or a street of decaying, once-imposing houses.

Film is a perfect visual medium for the coming together of place and emotional memory. The 1949 movie *The Third Man*, directed by Carol Reed, is a classic. Holly Martins, the protagonist, is not so much a hero or an anti-hero as a regular, war-naïve American guy, who makes his living churning out 'cowboy and Indian' novels. Through his eyes the camera follows the memories left by his dead friend around the streets of Vienna, as Martins tries to piece together past events. The music is brilliantly timeless. It is all mystery: we are lost in an irretrievable shadowy past, being guided by the streets of the city. Her streets have witnessed the extremities of human behaviour that the Second World War drew out, both selfless endurance and evil exploitation. The decaying city seems to embody the morbid memories of the war. Orson Welles appears fleetingly, suspiciously scanning the streets before disappearing into a doorway. Emotional memory is induced through the oblique cinematography of the buildings and doors on the seedy streets where the phantom past is staged. The denouement is a memory that is replayed on the streets, bringing the mystery to its grim conclusion.

So what is the magic resonance that is stirred by the Paris of *Afterimage* or the Vienna of *The Third Man*? Patrick Modiano and Carol Reed have entered the intuitive emotional memory systems of the reader/viewer, using the same pathways that are employed for autobiographical memory recall. Emotion becomes entangled with place memory through the amygdala–hippocampal connections. The emotional memory is knitted into the neural connections between the hippocampus and the amygdala so that seeing that place will subsequently fire an emotion. Neuroscientific discoveries have now advanced to the point where the very proteins that a rat makes in hippocampal cells following a fear response in the amygdala have been identified.[55] The magic of these works of art based in cities is that they intuitively find and fire these sensitized loci in one's individual brain.

I will finish with a buried place memory that was snagging at my attention when I wrote the first draft of this chapter. It was a fragmented memory of a 'short', a 3- to 5-minute movie. In my youth, shorts were usually shown before the main feature in arthouse cinemas, and it was in such a venue, the Lighthouse Cinema in Abbey Street in Dublin, that I thought that the event had happened. I had vague snapshots of the film . . . a man is in a barn and he is tinkering with a car engine under a raised hood, listening to the radio. The news that Elvis is dead comes over the airwaves. The barn is lit only by a lantern, so that when the man and his lover begin to dance they are dancing in a circle of warm light in the dark barn, like the lighting around the figures in a Rembrandt painting. I didn't remember the main feature that followed the short, nor who I was with, what the short was called, or the year. I went in search of the lost short and eventually tracked it down. It was called *That's All Right*, after the Elvis song that the characters dance to, and it was released in 1989. I also didn't remember that the late Mick Lally, a fêted Irish actor, was the star. I had only remembered places: Abbey Street, the interior of the Lighthouse Cinema, now gone . . . the lantern-lit barn . . . places within places within places . . . in this case the place where two individuals shared a universally remembered moment, their 'Where were you when Elvis died?' moment. Incidentally, *That's All Right* had been made for a tiny sum of money from the ends of a reel of film left over from John Huston's gem *The Dead*, based faithfully on the exquisitely poignant short story of the same name written by James Joyce in his collection *Dubliners*. These famous short stories are all set in the streets of Dublin and the homes of Dubliners. *The Dead*, perhaps coincidentally, perhaps not, in its haunting denouement set in Gertrude's childhood family home that houses her hidden heartbreak, makes us feel how the dead inhabit the living in their memories.

So . . . the magic resonance of place-evoked emotional memories goes on and on, and back and back . . . back to the earliest memories of the childhood home . . . deep down to your buried hippocampal place cells and their idiosyncratic tangles with your amygdala/insula neurons . . . your very own psychogeography. We can walk through a city and remember 'through its avenues, the feeling I'd once had of being light and carefree',* or see a small gravestone and feel terror. Places – the streets of Dublin, the avenues of Paris, the doorways of Vienna, Edith's gravestone, the herb garden in Cambridge, the childhood home, the haunted house, the barn dance of remembrance – are, experientally, the anchor for memory and feeling.

* Patrick Modiano, 'Flowers of Ruin', from *Suspended Sentences*.

7. Time and Experiencing Continuity

'Yesterday! What does that mean? Yesterday?', Hamm asks Clov in Samuel Beckett's *Endgame*. Hamm is raising one of the big questions about human memory – what is time and what does it mean? Physicists have been preoccupied with measuring place, space and motion for millennia and have more recently turned their attention to time – it is really their Big Question. They have long known that, while place and space can be measured as matter, time can only be measured in relation to place . . . a 'light year', for example, is a measure of the distance – not of the time – that light travels in one year. The units of time are fashioned around the momentum of the Earth within the bigger universe, specifically in the movement of the Earth and the stars. A day is the measure of the time that it takes for the Earth to rotate one complete revolution on its own axis, a year is the time that it takes for the Earth to rotate back to the same point in its circular movement around the Sun, and so on.

Time is more like momentum and happenings – like the light wave moving from the star, like the Earth hurtling forwards in a spin around the holding magnet of the Sun. 'Event physics' reflects the new realization that time is an intrinsic part of place and movement. The idea of an 'event' as the basis for lived memory has been around since Brenda Milner grappled with the mystery of Henry Molaison's memory loss in the 1950s. HM, without his hippocampi, had a memory for neither place *nor* time. From then on it was understood that place and time were memorized together as events. There is event memory and event physics; and concepts of time, place and momentum are fundamental to understanding both. In this chapter we will look at how time is clocked in a similar way in the brain as it is in the world of physics, in a dynamic pattern of shifting locations and events. In the previous chapter we followed the place cells to explore the psychogeography of emotional memory, and now we will explore how time is also tethered to the compass of place in the hippocampus.

Clocking Time

We often ask, 'Where has the time gone?' Nora, a patient who suffered from bipolar disorder, had good reason to ask this question when she was admitted to the hospital some years ago. Bipolar disorder is an illness in which the sufferer has episodes of depression and of mania. Along with a fathomless sadness, depression brings with it a sense of exhaustion and cognitive slowing, in which memory is impaired and thinking is muddled. Mania is on the opposite pole, with experiences of emotional euphoria and uncontrolled cognitive excitement. The case of Nora's experiences of time demonstrates the, sometimes extreme, effects that mood disorders can have on the sufferer's sense of time and also demonstrates how we all need some basic brain functions to clock time.

Nora had been unwell since early adulthood. She was admitted to an inpatient unit in a manic state. She was hyperactive and had been travelling the country, visiting people who she used to know but with whom she had lost contact over the years. She was talking in an uninterruptable loud voice about her grandiose and paranoid ideas. She had multiple delusions with a superficial logic that did not withstand much questioning. Nora would begin by telling her story in a chronologically disjointed way, and if the listener responded and listened for long enough she would progress to accusing them of being part of one or several vague plots against her. Following some weeks of escalating difficulties, Nora was brought to a psychiatric inpatient unit for involuntary treatment. She was known to the older nurses because years previously she had been in and out of the hospital frequently but had been apparently in remission for many years. (Remission means an absence of illness in a disease. A recurrence of an episode of illness is called a relapse.)

Things became clearer when it was realized that Nora had not been in remission but had been in a 'shut down' state for many years. She had not left the family home, had not read a newspaper or watched TV, and had spent most of her days in a quasi-wakeful and unresponsive state. Then, without any warning, some weeks before her hospitalization, she 'woke up' in a manic state. The most extraordinary aspect of Nora's presentation was that her references for the world dated back several years to the time when she fell into her psychic hibernation. During her missing years Ireland had undergone an economic boom, and the accelerated economic prosperity was evident in new building developments and in an updated public transport system. A major city centre street had become pedestrianized, the traffic flow had been altered and clothes and hair fashions had changed. Nora appeared not to have laid down memories for her years of depression. Her capacity to make new memories was not affected by her brain hibernation and, once she had struggled through the disorientation of the time warp, she gradually established herself in a fresh world.

Nora was in a state of under- or hypo-arousal during her hibernation. Neurologists call this an obtunded state, which describes a state of reduced awareness. This can happen in the sort of depression that is experienced in bipolar disorder, although very uncommonly for a prolonged period of years. Nora went from an obtunded state of depression to the hyper-aware state of mania. One of the reasons that Nora's story is so instructive is that, during her missing years, when she was in a state of blunted awareness, she had apparently laid down no memories. She was able to retrospectively fill the gaps of the missing years with news from the world, but her personal narrative memory, her own biography, did not contain lived experience for these years. Time had not existed for Nora because it had not been recorded.

Contrast Nora's experience with that of the fictional Miss Havisham, in Charles Dickens's novel *Great Expectations* (1861), who was abandoned by her fiancé on their wedding morning. She closed the doors of her big house, remaining in her wedding dress and veil, having stopped the clock at the hour when she was abandoned by her betrothed. The irony is that, try as Miss Havisham might, she could not make time stand still. Time did not stop for her even if the clock did . . . cobwebs grew, her dress frayed, her face and body aged, and the time that she refused to leave moved on. Her memory-formation continued, and she did not remain romantic, bewildered and amnesic, but as her hippocampus clocked up the years her personal narrative sclerosed and her emotions soured, preventing the possibility of any future that could have saved her. We may sometimes want to stop the clock, freeze a moment in time, pretend that 'it never happened', we may want to forget for the rest of our life that moment when everything changed – there was a before and the forever after – but time cannot go back and the present cannot remain the present. Events happen whether we want them to or not, but event memory will only be recorded if we are aware and conscious.[i] Arousal, awareness and consciousness are key concepts as to whether we memorize, and how we memorize.

Nora's experience may seem like an outlier compared to the average person's experience of time. But is it? Years ago I got food poisoning abroad and remember being very sick a few hours after eating some shellfish, and the next thing that I remember is coming to, to my surprise, one day later. The only thing that I had remembered from the lost day were a few moments of an undulating wall opposite the bed and knowing that I was delirious. Being awake and conscious are the first requirements for experiencing time. Nora was in a state of reduced awareness and did not experience the world around her like a normally wakeful, sensate person. Without awareness, or consciousness of the present, events cannot be recorded. The fictional Miss Havisham, being sensate and conscious, could only pretend that time had stopped, but she was alert and making memories.

Essentially, time is clocked through a recording of events. Consider the intellectual brilliance of the physicist James Clerk Maxwell when he described this intuition in 1876: 'The idea of Time in its most primitive form is probably the recognition of an order of sequence in our consciousness.'[ii] Since time can only be experienced if one is at a critical threshold of consciousness, let's look at consciousness and in so doing we may find an answer to Nora's lost time.

Exploring Consciousness

Exploring consciousness is a bit tricky, and if one has any tendencies to wander into the realms of spiritual explanations, consciousness is the most plausible way to enter these heavens – why do we continue to use the language of 'the soul' when we don't understand something in the brain? The word 'consciousness' is a problematic one because it is a catch-all that can mean anything from being awake to being highly aroused, to being in a state of transcendence, to seeing yourself as if you were another person looking at you, or imagining yourself as if you were someone else. Not only this, but you are conscious of your own consciousness and of the consciousness of others. So, what are we talking about when we talk about consciousness?

Freud is culturally associated with the concept of the conscious/unconscious. The unconscious mind, in a Freudian sense, contains fantasies and memories that have been put out of the reach of awareness, or conscious recollection. According to Freudian principles, while unconscious memories and fantasies influence responses to the world, this influence is out of the bounds of immediate consciousness. People with repressed memories are nevertheless driven to feel and act by memories of which they have no awareness, because non-conscious memories, similar to ones that we have awareness of, bring about emotions, and sometimes the emotions lead to unpredictable actions. These feelings and actions may not be consistent with an individual's beliefs about their narrativized selves, and so the individual feels at odds with themselves and the world. An example would be a fear of sexual intimacy, following repressed memories of sexual abuse during childhood. This could lead to a rejection of all forms of intimacy although the survivor of abuse, like all of us, wishes to be loved. As I've outlined in previous chapters, Freud theorized, in keeping with the permissive pedophilic sentiments of the late nineteenth and early twentieth centuries, that a child was attracted to their parent of the opposite sex.[iii] Girls were not only attracted to their fathers, but they were jealous of their penises! Although I tend to have an unreconstructed primitive amygdalar reaction to Freud's brand of misogyny, one of his more enduring and important legacies is the idea that memories that lead to feeling states are not always present at levels of conscious awareness.

The experts in pragmatism, the medics, have an entirely different view of consciousness, and theirs is the steadfast common denominator of all definitions of consciousness, however esoteric. In medicine, consciousness is measured by levels of wakefulness, often referred to clinically as 'arousal', and since there is some clarity about the physiological substrate of wakefulness and what this means, let's start there. We will call this level of consciousness 'arousal consciousness'. It is evident that without arousal to a state of wakefulness there can be no registration of events. One may be normally awake and aroused, or may have reduced arousal, for example when sleeping, or dozing or drowsy. Within the normal spectrum of wakefulness a person will better remember when moderately aroused. Outside of this normal spectrum there are pathological states of arousal. A person may be in a non-rousable state, unconscious, following a collision trauma or severe blood loss. A coma is a prolonged state of unconsciousness, and it can be measured in simple bedside-test scales, the most common one being the Glasgow Coma Scale. A score of 15 is normal, while one between 3 and 8 indicates a comatose state. Part of the routine examination of a patient is whether they are 'orientated by three' – that is, responding verbally to indicate that they can identify time, place and person. On the other end of the arousal spectrum a person may be highly aroused, for example if experiencing intense emotions or trauma, or acutely psychotic or intoxicated by a stimulant drug like cocaine. If this arousal is extreme, registration of events may not be possible.

The brain structures involved in the coordination of wakefulness start in the brain stem, situated between the spinal cord and the cortex, from where discrete circuits connect to different brain regions, particularly the sensory cortices. It seems that there is a gateway mechanism that controls the access of sensory information from the body to the cortex, and if this is shut – in states of impaired consciousness such as a coma – then the cortex will remain in a sleeping, low-voltage state of arousal. Cells need to be charged to fire and permit sensation and memory processes. If the brain stem 'wake up' switch is malfunctioning, cortical neurons will not wake up and will not fire and flare to create sensation or to form memories of the sensate world. When the brain stem is destroyed, a person is thought to be brain dead. Awareness consciousness is also dependent on the integrity of multiple long connecting tracts from the brain-stem switch to the various cortical areas.

We sometimes see states in psychiatry in which patients who are awake have some minimal wakefulness and arousal – for example Nora's shutdown period – and they will have commensurate degrees of amnesia for the periods of reduced awareness. We need first to be awake, and then have a requisite level of awareness, to make memories. A common example of a wakeful state where there is loss of awareness is being very intoxicated with alcohol, when there may even be rowdiness and superficially coherent speech but there is frequently amnesia for the events that have happened – the alcoholic 'blackout'. We often see pathologically aroused individuals in acute psychiatry – Nora's manic state following her years of a semi-stuporous one is an example – although we don't have routine clinical measures for such pathological, highly aroused states. In manic states, individuals have greatly reduced need for sleep and can remain hyper-aroused with only intermittent sleep for weeks, or even months if severely manic. We manage the illness by titrating increasing does of medications – antipsychotics, sedatives or lithium – against arousal until the individual reaches a normal state where the sleep desperately needed to settle the brain is possible.

Hibernation presents a naturalistic example of the cycle of sleeping and wakeful states. As a child, I loved stories about squirrels rushing around during autumn to gather and secrete away acorns for the long sleep of winter. This feeling was pleasantly rekindled when reading a paper recently about hibernating squirrels.[56] The authors looked at hippocampal function during the two states of the squirrel's annual cycle. They found that during hibernation there are changes in a squirrel's hippocampal activity. There is a reduced connectivity, and dendritic concentration falls away. During the switch from the big sleep of hibernation to the spring the process is reversed, and frenetic activity is seen in hippocampal dendrites. I imagine the pre- to post-waking shift in squirrel hippocampal activity as being something like the switch from depression to mania in bipolar mood disorder. We will explore the next level of complexity in consciousness, the representative forms of consciousness, in the next chapter.

Time Cells

Let's now look at how time is clocked and recorded in normal states of arousal consciousness. Think of what you have done in the last few days. You remember it in a time sequence, probably of days and certain times of those days, perhaps a bit jumbled, but even in this you know that you are jumbled. If you know that you can't remember the sequence of events then you have memory for time sequencing. This can be generalized to all forms of memory – knowing that you have forgotten something is a form of memory, even if very unsatisfactory when it happens non-stop over the age of 55 with glasses, mobile phones and keys. As you were thinking of the last few days, you were thinking of events – going places and being in different locations, meeting people, personal exchanges, having meals. One's sense of time is inseparable from events, but this is a *sense* of time. Might time have something to do with place cells?

In the middle of the twentieth century Donald Hebb, who worked out how neurons melded together to form cell assemblies, intuited that the hippocampus could record cell assemblies of place in some sort of sequence that would reflect what we call 'time'. Like the physicist Maxwell, he speculated that time was somehow integrated into place memory, rather than there being a separate time system. Experiments in the first decade of this century in rats demonstrated that *sequential* cell firing in rat hippocampal cells seemed to represent the temporal sequence of events experienced by Rat.[57] The key to understanding the modus operandi of this idea is that events are recorded in sequence, like the reel of an old film where each snapshot follows another in a forward-moving temporal sequence. The images are substantial, just as external place/space is, and it is their juxtaposition that gives a sense of time. Time exists in a movie because the images follow one another to create a sense of moving forward from the last snapshot – events continue to happen in a forward moving direction because they are recorded in this way. It is the momentum of the images that confers the sense of time.

The sequentially firing cells in the hippocampus were called 'time cells' and seemed to have similar properties to place cells. It now seems that the inputs of time and place cells are integrated into a unified cell assembly system in the hippocampus.[58,59] The timespace memory thus formed is, at its most fundamental level, an event memory.[60] Let's pause for a moment to consider this and return to the question that led the last chapter on 'place': where were you when Elvis died/when 9/11 happened? There is a where and a *when*, woven together experientially. This is what is processed in the timespace cell assembly in the neural-memory production line of the hippocampus. The momentous event is then projected to prefrontal autobiographical stores. The event, consolidated in biographical memory as a unit of connected cells, is also retrieved as a unit. Biographical memory recall seems, although this is not definitely proven, to involve the prefrontal biographical memory *together with* the video-directing properties of the hippocampus. It seems to me that old-old memories seem to lose the momentum of the film-like recording of recent events and become more like snapshots of a place from the past, with only an accompanying sense of when it occurred. We have more certainty about place as event memories age than we have about time. You will probably have experienced this uncertainty about the timing of events when reminiscing with family or friends. Someone remembers an event happening somewhere and there is a general 'Ah yes, I remember,' and then the 'When was it?' starts, and now it is a free for all . . . the calculations follow, matching the event in question to others that occurred before or after. In this familiar process of collective reminiscing we 'place' the event in time through juxtaposition with other events.

The Fluidity of the Experience of Time

One can only gasp at the sheer cleverness of it all. Space and time are being recorded together, just as we experience it, just as we have forever narrated it, just like the physicists have been telling us since the nineteenth century. The relativity of time with space is *built into* the process of hippocampal recording. This has been elegantly articulated by the neuroscientist Liliann Manning, in her reflective paper on the journey from philosophy to neuroscience in our understanding of time: 'Experienced continuity is possible only by and through memory.'[61] Manning's description of time as being 'experienced continuity' was written from the perspective of a neuroscientist in the twenty-first century, but it is remarkably similar to the intuition of the nineteenth-century physicist Maxwell that time is 'an order of sequence in our consciousness'.

There is a hidden complexity to Liliann Manning's description of time as 'experienced continuity', demonstrated nicely by a small encounter between Alice and the Queen in Lewis Carroll's *Through the Looking-Glass*. Alice is shocked by the experience of the Queen, whose 'memory works both ways', that is, both backwards and forwards, to which the Queen responds, 'It's a poor sort of memory that only works backwards.' We mostly think of memory as only working backwards, because we experience time as going forwards; but it is not the queen who is daft, it is Alice and the reader. Time seems to go in one direction, from the past to the present to the future, but is this really how we consciously experience it? At some point in your life you will have had the feeling that a memory of what you are experiencing right now will remain with you for the rest of life. This can happen when we have experienced intense emotion, either happiness or sadness, perhaps during your first romance or at your wedding, or following your child's birth. You can feel as if you are in the present and in the future. Or it may be a day like any other day that contains a perfect moment – a moment that rouses the deep sense of familiarity that is contentment. I recall feeling, as I experienced a beautiful hot day on the beach on Ireland's Eye, a small island off Howth, with my friends and our children, that I would in the future always remember that moment of that day. Maybe I decided to take an amygdalar-hippocampal timespace snapshot, but it is more likely that the moment chose me.

In such moments the experience involves the present and the future because you are travelling forward in your biographical life. I call this experience a 'prescient memory' – it is a consciousness of memory formation, a feeling that one will recall this moment for the rest of one's life. The term 'prescient memory' is one that is used in computational models of prediction in AI, but I use it here in an experiential context. Prescient memories occur during moments of intense experience of the present. There is a particular heightened consciousness of self, a slight hyper-perceptiveness, an awareness of oneself as a separate sensate being in time in the world and in your own memory. This familiar human feeling again reflects the principle that the more aware and attentive we are, the more likely we are to form lasting memories.

If you are a parent with adult children you will be familiar with the conscious sense of the past merging with the present when your watch your children, even if adults. When my 18-year-old son was leaving home to attend college in Galway, I had a rapid succession of visual memories as I waved him off. The milestones of his life ran through my mind, flitting images in quick succession, like an old-fashioned slide projector, illuminated in a background of a diffuse emotional tone: bringing him home in a white cotton blanket to our house in Cambridge following his birth . . . leaving him at the creche on my first day back to work . . . standing outside our front door in Howth with the seascape to his back, holding his younger sister's hand with their little lunch boxes on her first day at Montessori school . . . taking a photo of him on his first day in secondary school looking like a little boy beside his friend who had sprouted . . . and now his sad face as he left us. Where had all the time gone?

We can go one stage further – you can sometimes have the subjective experiences of time as simultaneously going forwards *and* backwards, with full awareness that you are in the present. I remember having this sense of time moving in both directions one evening when holidaying with my family in Vancouver. I jogged around the sea front with my brother in the early evening, and later we all rambled to a local restaurant, regressed with our children back to own childhoods. Our long-term memories were merging with the formation of our current memories: past and present came together and there was a sense of continuity into the future looking at our own children horsing around as we once had. During that evening time felt like it was proceeding in two directions – from the present to the past, and from the present to the future. We were back in time and yet in a conscious and present experience of a prescient memory. This is the familiar feeling that we have with old friends, who are the best, when we get a contented glowing feeling of the continuity of one's life lit from the past and projecting into a warm future.

In 1899 the philosopher Henri Bergson wrote that 'Time is invention and nothing else.' The fact is that the concept of time as commonly understood is unsuited to understanding what is happening in dynamic existence. Events happen and are recorded in biographical memory in sequence, and it is the recall of these events in some sort of sequence that forms a sense of time. Proust's search for lost time – *À la recherche du temps perdu* – was an immersion in his memory. It is clear that the past does not exist except in memory. Perhaps the popularity of the 'time machine' fantasy, in spite of it requiring a suspension of all reason and the rejection of existential experience, lies in the fantasy that the past is not lost.

The idea that there is no time as such, no past and no future, has existed since the fourth century, when St Augustine of Hippo identified three times: 'a time present of things past; a time present of things present; and a time present of things future.' Augustine seemed to be saying that only the present exists, from which position one slides either to the past or to the future. His contemplations were reflected fifteen centuries later in the work of Endel Tulving, a Canadian neuroscientist whose life's work has explored time and memory.[62] His influential work has looked at how the past and the future exist in the consciously experienced present. Daniel Schacter, who worked with Tulving, made big inroads into understanding the experience of time in his neuroimaging experiments. He showed that the *same* brain circuits are employed when thinking about the past and when planning for the future.[63] But of course this is not surprising when you consider that we make decisions about the future based on experience, on memory. We can only fantasize, or predict, with the experience that is woven into memory. Memory is more than a record of the past – it is also the template for an imagined future. The past–future fusion of neural circuitry involves multiple integrated areas and, in particular, the hub of the memory machine – the hippocampus – and the integrative storyteller – the prefrontal cortex.[63, 64]

This all means that the future, as well as the past, resides in memory circuitry. In Alzheimer's, the hippocampus becomes damaged early on, leading to the first symptoms of the disease, memory loss. Those with Alzheimer's lose their sense of time as the disease progresses. In one experiment, a group of individuals with mild Alzheimer's were compared to age-matched controls with normal cognitive function. Those with Alzheimer's had an equally impaired ability to remember past events *and* to envision future events, 'providing further evidence of the close linkages between the mental representations of past and future'.[65] In Alzheimer's disease, past memories, stored in the prefrontal cortex, remain relatively intact in the early stages of the disease, when the plastic hippocampus is more quickly destroyed, but the disease gradually destroys autobiographical memory in later stages of the illness when the frontal lobes become more eroded.

There is one more aspect of the fluidity of past and future time that is worthy of mention because it is a subjectively fascinating and universal human experience. Time seems to stand still during childhood, in fact it doesn't exist experientally. The time is all 'present', days seem endless and events just end and move on to the next one. Children are not so much adaptable as partly amnesic – we will be exploring this in Part 2. Dylan Thomas put this vividly in his poem 'Fern Hill', that lilting ballad of the hurtling sensory flow of childhood: 'Time held me green and dying / Though I sang in my chains like the sea'. As one gets older, the subjective sense of time gradually accelerates, and, as you move into older adulthood, time flies by.[iv]

Is Time Really Consciousness?

As we have seen, place is based in physical matter that is apparent through the senses and can be measured objectively, while memory may be the *only* yardstick of time. We have also established that the past and the future reside in memory circuitry. If time exists only in past memory or in future imaginings, what of the present?

The present has been the common fulcrum in ideas of time from St Augustine's musings in the fourth century to Edel Tulving's neuropsychology of the twentieth. Is the present the real 'time'? The whole concept of time is generally unhelpful in understanding science, be it physics or neuroscience. It is more coherent conceptually to shift 'the present' away altogether from a time concept and reassign it to the concept of consciousness. From the perspective of the recording of events, the present *is* consciousness. In a seemingly ironic twist, I myself think that the only place that time does not exist is in the moment of consciousness. The past and the future are more like what we conceive as 'time', but the present belongs to consciousness.

The intellectual convergence between the disciplines of physics and neuroscience in our understanding of time and memory is truly fascinating, but memory is less like big physics, and more like particle physics, in which precise localization of an event cannot be achieved because of inherent uncertainty, a concept called indeterminism. Just as tiny matter becomes less predictable as it is removed from the holding force of gravity, memory becomes less certain as it recedes from the relative certainty of present consciousness. Events continue to happen and the dendrites continue to rearrange, making the original memory less exact. The restless neural activity, the non-stop momentum of the electrochemical and matter exchange in the neuron and among neurons, the growing and fading entanglements of dendritic arborizations as the neurons flare one into the next and the signal fizzles, is neverending. There is an infinite possibility of direction for our 68 billion neurons from the constant input of exteroceptive sensation from the world and interoceptive sensation from the body. The only certainty is that there will continue to be events that will reshape the cell assemblies. Our brain, like the universe, is a cauldron of entropy.

Sean Carroll, a contemporary physicist and writer, brings neuroscience and physics together succinctly: 'time cannot be measured in physics, unlike place or matter, it can only be understood subjectively.' In his book *The Big Picture: On the Origins of Life, Meaning and the Universe Itself*,[v] he looks at time as experience rather than as a potential concept of the out-there physical world. But for me there is an even more fundamental point in Sean Carroll's observation about subjective understanding of time. We understand ourselves and others, the world and the universe, through the only system that we have available to us – the cellular and network organization of memory. The neurons that make memory – the hippocampal, amygdalar and cortical cells – have a four-sided shape like a pyramid – four triangular surfaces – and, for this reason, are called pyramidal cells. Is it a coincidence that the timespace concept is four-dimensional, with the three dimensions of space interrelated to the fourth dimension of time?[vi] Maybe the timespace understanding of physics is based on the only way that we can see or learn about patterns in the physical world through our four-sided pyramidal neurons that interpret and memorize three-dimensional space and time as integrated events. Are the physicists – in their elegantly reduced science – in fact looking at how we memorize? This speculation will, if nothing else, help you to remember the four-sided pyramidal neurons beavering away and reflecting the four-dimensional timespace of life in motion. It is staggering that the organ that psychiatrists tend to and mend, containing timespace cells, embodying our individual narratives, working beyond the most sophisticated understandings of the physical world, is often not considered physical.

Let's finish our discussion where we began, with Hamm asking Clov what 'yesterday' means. A scientist might say that it is the events that occurred between now and the preceding full rotation of the Earth on its own axis, and a neuroscientist could say that it is the difference in the dendritic arborizations caused by the events that occurred during this rotation. But maybe the important question is not what yesterday or tomorrow means, but what the present means, because it is this that will determine the memories of past events and the fantasies and strategies for projects to come. Conscious experience, where time does not really seem experientially to exist, is where we make the past and shape the direction of events that have not yet happened. As St Augustine wrote, and as the physicists are on the verge of explaining to us, 'we measure time as it passes', and it is only in memory that our *sense* of time exists. But it is also only in the present where we can really experience the world and that is the gift of consciousness, of being alive. We have to cram both memory and sensory awareness into the conscious moment, and in all of this 'time past and time future' it is only too easy to miss what St Augustine called the 'present of the present or attention', the glorious sensate *now* of the conscious moment. Or as Pooh Bear tells children, 'Yesterday is history, tomorrow is a mystery, but today is a gift. That is why we call it the present'.

8. Stress: Remembering and 'Forgetting'

There is no state of mind, however simple, which does not change every moment.

Henri Bergson (1859–1941)

As established in the preceding chapter, we need to be conscious to the level of being awake to record the present, and to make memories. This means that the brain-stem switch for wakefulness needs to be on, and neural activity in the hippocampal and cortical neurons needs to be at a requisite level of arousal. The level of arousal of recording neurons is fundamental to what is memorized. What we are aroused by is individual and varies for each event within any individual's life, but is dependent on the same mechanisms in all of us.

Arousal involves the two hypothalamic exit systems – the ANS and the cortisol stress system. This means that your neurons are aroused when you feel aroused, which makes it easy to establish whether you are recording and can memorize or not. I was once very excited by the science of stress and the relationship of stress to depression – it has formed the bulk of my life's research work. During the 1980s and 1990s there was limited interest in stress research, and we were quite a small group. The research discipline was called psychoneuroendocrinology, and it has remained as a specialist research discipline forming a foundational knowledge-base that has now been applied to most diseases. Scientific stress literature originated in endocrinology, because cortisol is a hormone, and in psychiatry, because severe brain stress is present in mental illnesses and may be germane to the cause of depression. Back then, the idea that the stress hormone cortisol is a proxy measure of physiological and psychological stress, which is now common knowledge, had not yet been established.

In this chapter we will look at the effects of stress on memory, best understood through the pathology of depressive illness. In 2003 I had the good fortune to debate on the same side as Lewis Wolpert in one of the Maudsley Series of public debates. We were opposing the motion, 'This house believes that antidepressants cause dependence'. Lewis is better known for his crystal-clear writings on science than for his experience of depression, but his book *Malignant Sadness: The Anatomy of Depression* focused his scientific brilliance on the topic of depression and is well worth reading. I remember Lewis telling the audience that his depression had been a worse experience for him than the death of his beloved wife. He found, as most people who have experienced depression, that one of the most disabling aspects of this grim experience is that it scrambles your memory and your thinking. This is often the reason that people present for treatment: they may struggle on through the despair and self-loathing, the fatigue and the sleeplessness, even the suicidal urges, but they seek treatment when they find themselves unable to concentrate and memorize, and unable to work.

I will now tell you about Sally, who was a patient of mine.

Sally

A colleague from the medical wards asked me to give an opinion on Sally, who had been admitted from the ER a few days previously with an abrupt and generalized deterioration in her health. She had not left her bed at home for about a week and had stopped eating and drinking. She had stopped responding to her family and had gradually fallen into a mute state, and then a comatose one. I interviewed Sally's husband, who gave a clear history of normal function up until a few days prior to her hospitalization, although he remarked that she had been quieter and less active than usual in the preceding weeks. Sally appeared to have a stable emotional and personal life. She had a history of depression, having become depressed for the first time five years previously. During her first episode she had become withdrawn socially and fatigued, and was non-communicative, but she made a full recovery. She had remained on antidepressants for some years following the first depression and had stopped the medication about four months prior to her current presentation.

Since admission to the hospital, she had deteriorated further, and this deterioration seemed to be rapidly progressive. She eventually fell into a comatose state and was not responding even to mildly painful stimuli, such as a light pinprick. The medical team had conducted multiple laboratory and radiological investigations but nothing abnormal was discovered. What was strange was that, although her vital signs, her blood pressure, her pulse and even her temperature were in a higher range, and fluctuating in an unusual way, there seemed to be no medical illness. She was aroused physiologically – her ANS was overactive – but she was also comatose. ANS arousal usually gives rise to a feeling of being generally aroused. There was some major pathology at play but it could not be identified. Her clinical picture suggested the possibility of an encephalitis, an infection of brain tissue, but an electrical recording from her scalp (EEG: electroencephalogram) indicated low-voltage but normal electrical brain activity. Her brain imaging scan was also normal. A lumbar puncture yielded spinal cord fluid that had no detectable signs of infection or any immune disorder. There was no other source of infection that could be identified. She had very high cortisol levels, but this was considered a non-specific finding because she was so ill.

We stood at the end of the bed looking at Sally. She was lying on her back, motionless, eyes shut, with her lips slightly parted. She was occupying the bed closest to the nurses' station because she required fifteen-minute observations of her vital signs, and they reported that she did not seem to have moved for the preceding twenty-four hours. Sally was being hydrated with an IV line and had not had food for many days. I held her hand and spoke to her: there was no response to a gentle squeezing of her hand. I held her elbow with my other hand and lifted her arm from the bed, and then slowly disengaged my hands. Her arm remained aloft and did not move downwards for a few seconds, and then slowly descended to rest tensely on *the bed. Looking closely at Sally's face I saw a salt track coming from her eye, down the curve between the nose and face, to her upper lip. We had our diagnosis: this was catatonia.*

Catatonia refers to abnormal motor activity, specifically in movement and speech. There are particular movements that identify catatonia, one of which is the maintenance of unusual or uncomfortable postures. Sally's arm remaining suspended in an awkward position is an example of this. This maintenance of odd positions is known as 'waxy flexibility'. Another clinical sign of catatonia, related to uncomfortable postures, is called the 'psychological pillow': here, the sufferer keeps their head not resting on the pillow but raised slightly off it. It is difficult to imagine the extreme muscular tension and the mental suffering that is present in catatonia. Sally's abrupt and extreme form of catatonia is rare. It usually evolves more slowly from an untreated, or under-treated, mood disorder. The proposition that catatonia is the human equivalent of the animal state of being frozen in fear resonates with psychiatrists. Being 'scared stiff' may be a mechanism to fool a predator by feigning death and may be a hardwired behaviour that is generally buried in the human's more elaborate soft wiring of experience.[66]

The treatment for catatonia is benzodiazepines – Valium-like drugs – in high doses. This brings down the level of arousal and seems to unfreeze the catatonic state. We administered diazepam intravenously through a small bag of saline, and one hour later Sally came to and later that night had tea and toast. We were then able to administer benzodiazepines orally, and we recommenced her antidepressants. She was fully mobile but very slowed down within twenty-four hours and was discharged from the hospital some days later. I spoke to Sally in more detail when she recovered from her trauma. She had no memory of the days before her hospitalization or of most of her inpatient stay. She remembered her head had been in a spin for a few weeks before the onset of her amnesia, and her last memories before this were of feeling down and anxious.

Various degrees of amnesia are experienced during severe depression, usually proportional to the severity of the depression.

This amnesia extends from poor biographical memory to complete amnesia and can be seen in the poor general biographical memory in those who suffer from recurrent depression. Day-to-day memory difficulties are apparent to sufferers during the most rudimentary tasks: even following a movie plot or reading the newspaper can become impossible. An inkling of this experience that is probably familiar to you is if you have been so stressed that you are not able to think and you keep forgetting things. The amnesia is not caused by the drugs given to treat the depression, as is sometimes proposed, because memory actually returns following administration of medication. A commonly cited experience of stress-induced amnesia is when someone is given bad news by a doctor. Most individuals can remember very little of what the doctor has said about the disease or the management plan. Being aware of this inability to memorize when in an emotionally shocked state is a basic principle in learning how to break bad news.

How does this amnesia fit in with the idea that we have already established, of higher arousal or better attention leading to better memory formation? Should it not be the case that the more aroused a person is the more they should theoretically memorize? Memory, like many physiological systems, works best at moderate levels of arousal, and at the extremes of arousal, either hypo- (low), or hyper- (high) levels of arousal, memory function is impaired. A very loose rule of thumb is that low levels of cortisol cause low levels of arousal and poor memory formation, but high levels of cortisol causing high levels of arousal will also impede memory formation. Learning is thus less efficient at low or high levels of cortisol, and most likely to occur with a moderate level of arousal.[67] We know intuitively that we cannot learn if we are not paying attention or, at the other extreme, if we are over-aroused and anxious.[68]

Nora, in the previous chapter, lost years of memorizing the world because she was so under-aroused that she registered almost nothing. Sally's story, on the other hand, demonstrates how severe arousal can also cause amnesia. Sally's autonomic nervous system (ANS) had been in hyperdrive, and her cortisol levels were very high. She had been locked inside her body, as I imagine the experience, paralysed with fear. Over the last few decades it has been established that depression is associated with high levels of cortisol and that this is more pronounced in more severe forms of depression. To understand Sally's and Nora's experiences we need to look at the cortisol stress system, specifically the effects of the stress hormone cortisol on neuronal activity – what I and my clinical science pals call 'brain stress'.

The Hypothalamic–Pituitary–Adrenal Axis

The literature on the physiology of stress goes back to the origins of medicine as a discipline, to Hippocrates, who lived and died four centuries BC. The first theoretical suggestion of a stress system came from this great philosopher and medical pioneer whose altruistic principles remain the foundation principles of modern medicine. These principals are based in his belief that there is a system in the human body that is at work all the time keeping us well and which becomes more active to counteract illness and disease. This is what we now know as the stress system. An important doctrine of Hippocrates was maintenance of health, which can be translated into modern medical-speak as 'reduction of excessive stress' – a concept that we have only worked our way to in the last century.

We seem to have lost the foundational Hippocratic idea that some level of stress is necessary for healthy functioning, that stress is good, and the word 'stress' now is almost synonymous with illness caused by excessive stress. Stress is not just good but necessary for sustaining life, and it is only in chronic, or long-term, overexposure to stress that it can be destructive.[69] In the 1990s I led a series of studies in chronic fatigue syndrome, a difficult-to-diagnose condition characterized by body and brain – called 'central' – fatigue. We found that cortisol levels were lower than average in this disorder – in contrast to the opposite findings in depression, where cortisol levels were higher than average.[70] I think of cortisol more as an activating, rather than a stress, hormone, one that keeps us alive and alert.

The system that Hippocrates intuited was first identified physiologically over 2,000 years later, by Hans Seyle. In the 1930s, Seyle first described the body's stress system and identified cortisol as the body's main stress molecule. The stress system is ultimately controlled by the brain and is called the hypothalamic–pituitary–adrenal (HPA) axis, in reference to the brain centre and the body organs that are involved in the control of cortisol release. The hypothalamus – the headquarters of the autonomic nervous system that makes the rainbow of our emotional states – makes a hormone (CRH) that is carried in the blood system from the brain. This brain hormone ultimately causes the release of cortisol from the adrenal glands.[i] As we have explored in a previous chapter, memory pathways can activate the ANS, leading to emotional experiences. The web of inputs to the hypothalamus has been sculpted by the dendritic growths of memory, and in this way memories, built into brain networks, can also cause cortisol stress responses through the secretion of CRH that will go on to activate the HPA axis.

Cortisol secretion into the blood system is controlled by the brain and, once released from the adrenal glands, is carried to all organs in the body, including the brain. The most important finding in relation to understanding the effects of cortisol on brain function and memory was the discovery in 1968 of cortisol receptors in the hippocampus.[71] The late Bruce McEwan, who co-authored the landmark paper showing cortisol receptors in the rat hippocampus, went on to lead a stress research group in the Rockefeller University in Manhattan, a hive of productivity during these early exciting years.[ii] But it was many years before the brain-stress research community understood the significance of this discovery. For this part of the story we need to return to Leiden – the intellectual haven for the Sensationalists during the religious persecutions of the seventeenth and eighteenth centuries – to Ron de Kloet's lab. De Kloet is Academy Professor of the Royal Netherlands Academy of Arts and Sciences. His work demonstrates how excellent science is about paring down, as far as possible, a single question and finding the answer through staged experiments that adhere to the central question, until the answer to that question is revealed in a coherent and undeniable simplicity. One of the most difficult tasks is adhering to the central question and not moving off on one of the many interesting tangents that present themselves.

Ron looked at how cortisol works in individual neurons in the rat hippocampus.[72] Neurons have to be charged to a minimum level to transmit a current to an adjacent neuron. A critical period of neuronal arousal is needed to achieve the requisite electrochemical energy to form the proteins required for new synapses that make the cell assemblies that constitute a memory. Ron demonstrated that cortisol fires neurons for the period of time required for dendritic growth.[73] This provides an explanation of why memory cannot be coded if levels of cortisol are below a certain threshold. The neuron, like the brain, is sleeping, or maybe dozing . . . remember that whatever is happening globally in the brain is happening at a cellular level ultimately. A minimum threshold of cortisol is necessary for memory formation, and we will call this 'good' stress.

If, conversely, cortisol levels are constantly high, then the hippocampal neurons, like the catatonic patient, are in a constant state of over-arousal. You might think that this would result in super memory formation, but that is not the case, because a neuron has to return to a lower level of electrical activity before it can be recharged to form another memory. The over-aroused neuron is stuck in a hyper-fired state and is unavailable for new stimulation. This is why unremitting high levels of cortisol, or 'bad' stress, inhibit memory formation. The same is true for ANS arousal, also needed to fire up neurons: low levels of noradrenaline will not fire neurons adequately, while high levels will not allow the neuron to baseline and recharge, that is, it will be refractory to charge.

I will now tell you about another patient whom I treated some years after Sally, when Ron was publishing his animal work on the inhibitory effects of high cortisol levels on hippocampal firing. I was working as a psychiatrist in a Dublin hospital, following the return of my family from Cambridge, when I was asked to see a young man, Daniel. It was very exciting to find that I could translate Ron's laboratory findings to real-life psychiatry.

Daniel

Daniel was admitted under the Endocrinology team. His family doctor had found, during routine blood tests, that Daniel had a raised blood glucose level and new-onset hypertension. His metabolic instability was progressing rapidly, a very rare occurrence in a young man, and he was admitted to the hospital for investigation. The endocrinologist suspected that he may have a cortisol-secreting tumor, because cortisol also controls the release of glucose to the blood circulation – acting as a physiological antagonist to insulin – and laboratory findings demonstrated very high cortisol levels, consistent with a possible tumor. His family gave a history of escalating strangeness in his behaviour and increasing remoteness over quite a short period of time, a few weeks at most. His mental state was rapidly deteriorating, and I was asked for advice in relation to this aspect of his care.

Daniel sat at the edge of the bed rocking to and fro. His hands were clasped in front of him, as if he was holding some imaginary object. I gathered after a few minutes that the phantom object was a rose. He was repeating '. . . smell the rose, rose, rose . . . smell the rose, rose, rose . . .'. When I spoke to him, he would repeat the last word of my sentence over and over. Once or twice his eyes flickered when I passed in the line of his vision, but he seemed generally oblivious to the psychiatry and endocrinology team assembled around his bed. He was staring past us all. I asked him to stand up, and after repeated requests he did so very slowly and stopped in an awkward semi-erect position, standing absolutely still and staring, for several minutes.

Daniel was in a classical catatonic state. In addition to the strange, uncomfortable postures, and the immobility, patients with catatonia may repeat words or repeat the last word of your sentence. Usually, individuals in this extreme state are unresponsive to other people, and even if they make eye contact it is glazed and unfocused. Daniel, contrary to what was expected, did not have a cortisol-secreting tumor, but had extremely high cortisol levels because he was suffering from a catatonic depression. We treated him with benzodiazepines and mood-stabilizing medication, and he improved within days, returning to his normal state within a few weeks. As his mental state returned to normal his cortisol levels reduced and his metabolic systems stabilized. His blood pressure returned to normal and healthy glucose regulation returned. Daniel remembered nothing from his admission to the hospital to a few days following commencement of treatment. His case confirmed what I had suspected about other cases of catatonia that I had seen. Individuals who were catatonic, although mute and/or immobile, were in a highly aroused state of ANS and cortisol overdrive.

Bad Stress and the Brain

Although there have been, and continue to be, many books written about managing stress, many of the stressors that people are exposed to are unremitting – for example poverty, socioeconomic deprivation, violence, racism. Life can be brutally hard on some, especially those born into socioeconomic disadvantage or exposed to abuse during childhood. The scientific literature on the negative effects of early-life adversity on the body, including the brain, has grown in tandem with the stress literature. There is now evidence, including my own research, that this process may start during fetal life.[74] Academic reports from the Rockefeller Institute published in the 1980s showed how rat pups who were shown affection – licking and grooming by Mother Rat – had calmer responses and lower cortisol stress outputs when exposed to a severe stressor in adulthood, compared to pups exposed to low mothering.[75] These findings showed how important very early events are in determining subsequent stress responses, and therefore general health. It emerged that cells in the hippocampus lost their dendritic complexity and synaptic connections fell away if Rat Pup was exposed to constant high levels of cortisol.[76] Bruce McEwan wrote quite beautifully of this process as being 'a dynamic brain architecture that can be modified by experience'.[77] The 'biological imprinting' of childhood neglect in humans, both in abnormal stress responses and in the hippocampal dendritic richness, is now well established.[78]

Whatever the cause of the chronic brain stress – childhood adversity, ongoing socioeconomic stress, genetically driven skewed stress responses, major mental illness – the hippocampus will suffer the toxic effects of chronic cortisol exposure, and memory systems will be damaged.[79] I have already mentioned the visible effects of hippocampal damage in human neuroimaging research in depression. The international collaboration ENIGMA (Enhancing Neuro-Imaging Genetics through Meta-Analysis), which brings together research from fifteen centres worldwide, including neuroimaging data from 1,728 depressed patients and 7,199 healthy individuals, has provided the most indisputable evidence to date that the hippocampus is damaged, or visibly shrunken, in human depression.[19] In our group we have found, in common with other groups, that it is the left hippocampus that is particularly shrunken and that the damage seems to be confined to the layers where the memory-making process occurs.[20] Depression, if untreated, is now known to be a disease that damages the memory production line in the hippocampus.

This all sounds very gloomy – childhood trauma seems prophetic of adult brain disorder. But all is not lost because, although fragile and relatively easily damaged, the hippocampus is plastic and can be repaired. There is increasing evidence that antidepressants and talking therapies can help restore a fecund dendritic interconnectivity in the hippocampus. Agents such as antidepressants may be able to cause 'reversible remodelling' of neuronal connections in the hippocampus, prefrontal cortex, and amygdala.[80] This happy finding that dendritic atrophy may be reversed can give us all hope about the potential therapeutic effects of pharmacological and psychological treatments on the brain.

People often say that they can't remember something. They have 'forgotten' events that others, who may have shared the experience, remember. This discrepancy in individual experience probably happens because the event was of no real significance to one person but may have had emotional resonance for the other. If we are not paying attention we do not mount the arousal required to memorize. One of the most densely investigated areas in neuroscience is why we pay attention to certain cues and are uninterested in others. Neuroscientists call this selective attention 'salience'. Arousal and attention provide mechanisms whereby we can remember individually important, or salient, information and discard more trivial sensory input. We are interested in things that 'excite' our curiosity, neurally speaking. In research led by my collaborator Thomas Frodl, we found that those suffering from depression had abnormal coupling of attention to emotional stimuli.[81]

Most of my friends are over fifty-five, and they often ask me, 'Why do I forget things?' If we can learn any lessons from Nora, Sally and Daniel, the answer is that they probably did not *forget* something: they never laid down a memory for it in the first place. Our memory systems will not function below, or above, a certain level of arousal. If the question 'Why do I forget things?' sounds familiar to you, you can first of all reassure yourself that remembering that you have forgotten something is a form of memory. Next, you need to question whether you registered the event. Activities that get one's attention, and paying attention, will increase lagging cortisol levels and improve registration of information. In common with other groups, we have found that exercising will stimulate arousal and improve memory function, and may even grow the hippocampus.[82, 83] Or, your hippocampal neurons may be in a state of fixed overdrive because of toxic stress. In this case, a therapeutic holiday in the sun may be the only cure. The HPA axis, like all physiological systems, can only work effectively to bring the body into equilibrium within a range of conditions determined sometimes by the environment and sometimes by the individual.

To Summarise Part 1

So far, I have torn a pathway through an eclectic mix, starting with the origins of neuroscience in the philosophy of the Sensationalists, and finishing with the recent convergence of neuroscience and physics in concepts of time and memory. The theme followed was that of experience, starting with the trials of the philosophers way back as they moved from divine explanations of Knowledge to the imprinting of knowledge from the world through sensation. We have followed the neural rush of sensory experience from the exteroceptive world and the interoceptive body as it gets woven into hippocampal and amygdalar memory, and is then consolidated, or not, in cortical memory. We have explored how neural input is organized into cell assemblies that will fire as a unit, and how these units connect up to allow an understanding of the world to develop. We have followed the oldest and most intuitive structure for recalling event memory – the time-place-person format of the story – to the depths of the hippocampus, to look at how this organic memory machine records time and place. Place is the solid thing around which the moving film reel of life events are recorded. As in physics, we record three-dimensional space with an intrinsic dimension of time, in the four-dimensional architecture of the hippocampal and cortical cells. We have followed the experience of arousal and awareness – mediated by the ANS emotional and the HPA stress systems – from the level of the neuron, where electrochemical energy is required for the formation of dendritic projections, to individual salience, or what gets one's attention.

Memories are made in the non-stop neural buzzing that is the living brain. Sensory signals come at us headlong, flaring the neurons like lights on a christmas tree, flashing on and off and in all directions, creating in each human an individual 'www' representation of their world in their cortex. An important point here is that there is no agency involved in where the sensory currents go, because signals are simply taken where the neuron assemblies have been laid down in the web of memory lattices. Does this mean that human experience is the product only of memory, that the particular memory maps of the neurons in any given brain will go on and on to interpret the world in an increasingly deterministic way? Yes, in the sense that we only have the network fashioned from the basic hardware and from previous experience; but this infinitely complex network of 68 billion neurons is in a dynamic, never-ending state of change. New experience arriving from the world through our senses will follow the 'best match', but will also forge new connections leading to an alteration of cell-assembly pathways. There is arrangement and rearrangement, lattices arborizing and being disrupted, memories being woven, being augmented, and being disassembled. Yes, humans are a construct of their own memories, but these memory systems are, at any moment of consciousness, in a dynamic never-to-be-repeated balance of the entropy of the incoming world making its neural impression within the relative stability of cortical memory maps.

In this neural momentum an individual lays down fluid neural pathways woven from experience and, somehow or other, forms a narrative framework that becomes 'them'. Let us now look at how this narrative framework is fashioned from the neural chaos . . . the story of how memory makes us.

PART TWO

How Memory Makes Us

The neural networks in our brain change from infancy to old age. Ideally suited to sensation gathering and discrimination, the infant brain develops into an organ adapted for abstract reasoning. The knowledge we imbibe and the events we experience become interwoven in the dendritic tangles of the ever-changing brain, shaping our sense of self and identity. We will explore how current experience constantly takes apart and reconstructs memory. A culture's collective memory ensures both an ever-expanding knowledge base and a sense of a shared past. In the final chapter I sort through some of our oldest collective memories as passed down in fairy stories, as well as through my childhood steeped in folklore and the stories of my patients, who taught me that the real matter of memory is experience.

9. Self-Recognition: The Start of Autobiographical Memory

Our memory is our coherence, our reason, our feeling, even our action. Without it we are nothing.

Luis Buñuel*

Once Below a Time . . .

Although humans learn so much more during infancy compared to any other period of life, very little from this period can be recalled. They acquire a vast amount of knowledge that becomes automatic, or implicit, such as learning language or walking, but they do not seem to have narrative memory. What is your first real memory? What age were you: two, three, perhaps four? For me it is sitting on a box looking down at an orange hand-knitted cardigan with a translucent orange button. I have retrospectively assigned the memory to our first house move, when I was almost three years old – the first memory of my mother's hand-knitted woollen cardigans that went on to mark successive stages of my childhood. This snapshot memory is typical of a first memory, which is usually recalled as a simple sense of oneself somewhere.

Before the orange-button moment my world is prehistory. Infantile amnesia is universal, whatever claims an individual may make. The coordinates of any story, as we've established, including one's own life, are time, place and person. 'Person' cannot be remembered until a person, starting with oneself, has been registered as existing. I have assigned the first moment of registration of *my*self to

* Luis Buñuel, *My Last Sigh* (University of Minnesota Press, 2003).

the orange-button moment. It seems to me that one's first conscious memory is one's first moment of self-recognition, or self-awareness. One has to see oneself before memory of oneself can begin.

In this part of the book we will look at how awareness of self develops in the brain and how this allows one's life to be memorized, and subsequently organized into a story of sorts. Let's start at the beginning of personhood with how humans, and some few animals, develop awareness of self.

Evolving Consciousness of Self

When I write at home I am often cheered up by my garden robin making his presence apparent with his chirrups. He has attention consciousness, he is bright as daylight and he is alert, and he has memory because, luckily for me, he returns to my back garden, as the constitutionally plucky defender of his memorized territory. But he is unlikely to have any form of self-recognition. By extrapolation we intuit that he does not have higher forms of consciousness, such as being able to consciously experience emotions, or to see himself in action fending off potential invaders with his singing, or to see other birds as possessing consciousness. Although Robin almost certainly does not have consciousness of himself, and therefore conscious biographical memory, he has a rudimentary consciousness of other robins, because he wards them off from trespassing in his territory. Robin needs to be aware of other birds, but not of himself, in order to survive.

A useful way of understanding the development of a human trait, such as self-awareness, is through evolution of this trait in older species of animals, a process known as phylogenetics. A rule of thumb is that human development from embryonic life onwards follows evolutionary development.[84] [i] Robin's working memory, in phylogenetic terms, remains at the level of the preconscious infant. In humans, as you might predict from phylogenetic principles, and from Robin, awareness of others is present before awareness of oneself. At about 6–7 months, babies begin to understand that their parent is an-*other* person, a person separate from them, and they begin to experience their first anxieties of being separated. Distress on being separated from the parent begins to show at this age because Baby is feeling this sense of being separate. Baby has to be coaxed gently from the unitary bond with Mother that starts during fetal life – 'the yolk and white of the one shell'* – to being independent. If Baby is left alone too long, or if they are not soothed to reduce the anxiety, or if they are of a very sensitive disposition, difficulty in forming trustful relationships – what is called 'insecure attachment' – may start at this point, leading to lifelong patterns of emotionally insecure attachments to other people.

About eighteen months later Baby begins to become aware of themself. A rudimentary form of self-awareness – visual self-recognition – can be evidenced by a simple test called the mirror recognition test, now accepted as being a proxy measure for basic self-recognition.[85] The test involves the application of a patch of rouge to the nose, after which the participant, an animal or an infant, is placed in front of a mirror. The rouge spot on the nose cannot be seen except through reflection. If an animal or infant reaches to their own nose to wipe away the rouge it implies that they recognize the reflection in the mirror as being a *representation*, but not the substance, of their self. They have passed the test and are considered to have self-recognition. The infant human has a sense that they are in their private interospective world and not in the mirror. If, on the other hand, they reach to their image in the mirror to wipe the rouge away they are considered not to possess self-recognition, that is, the mirror image, rather than being a reflection of themself, is them. Human infants learn, over time and in the following step by step sequence, that, first, what they see in a mirror is not another infant, second, that the mirror mimics all the infant's own motor movements, third, that the mirror does not have someone else behind it, and finally, that they are looking at a reflection of themself.

* W. B. Yeats, 'Among School Children'.

If you have a dog, you will know that they don't get beyond the first step. They don't know that their own reflection in a glass door is not themself and may, like our very sociable dog Nelly, scrape their paw on the glass back door to get out and play with the reflected dog who appears to be in the garden. Nelly, unable to recognize herself in her reflection, will bound out enthusiastically to play with the other dog. She looks around surprised at the absence of another dog and, after a few moments with her ears and tail cocked, goes off nonplussed to do something else. In spite of her highly tuned other-awareness, and her sensitivity to our feelings and even our intentions, she does not have this form of self-recognition. She cannot learn it, either: she simply does not have the neural machinery.

A limited number of animals do pass the mirror awareness test, for example orangutans, bonobos, chimpanzees, elephants and big sea mammals such as whales and dolphins.[86] These mammals are considered to have advanced forms of self-awareness, and we give them the respect that they are considered to merit because they are more like us. More recently, magpies when placed in front of a mirror were shown to pick away at a red spot put on their necks.[87] In order to remove it they try to peck the spot with their beaks and with their claws. Maybe it is not that surprising when you consider how clever magpies are, and how they tend to wait until you are out of sight before they get up to mischief. The magpie discovery is an indication that self-recognition – the basis of autobiography and higher consciousness – is not exclusive to humans or even to our immediate evolutionary ancestors.

Separation from Others

In *The Myth of Sisyphus*, Albert Camus writes:

> If I were a tree among trees, a cat among animals, this life would
> have a meaning or rather this problem would not arise, for I should

belong to this world. I should be this world to which I am now opposed by my whole consciousness . . . And what constitutes the basis of that conflict, or that break between that world and my mind, but the awareness of it?

Camus was, like many existential philosophers of the last century, grappling with the concept of self-awareness and consciousness. We are none of us *a tree among trees* or *a cat among animals*. We intuit *a break between that world and [our] mind*. The world is out there, and you, the person, are behind the door of the exteroceptive world in your private interoceptive world, looking out. You see yourself as separate not only from the world but from other persons. It may sound obvious that if you feel and think of yourself as just you and no one else, then others have to be separate, but it is also a profound point. It is only when this is absent, as it is sometimes in psychotic disorders, that one can see that self-awareness, and the consequent awareness of separation from others, is a process in itself. Most of us automatically know this at an experiential level. You know that when you think, do or feel something, that it is you who have had that thought, experienced that feeling or done that act. Your thoughts or feelings may have been influenced by many things, but you know that they originated in your own head and your own body.

Schizophrenia is an illness in which the affected individual frequently feels that their sensations, emotions, and even actions are not their own. Because experiences seem to not come from oneself, they are usually attributed to someone or something else. Awareness of self is so fundamental to human experience that you may consider it to be inevitable and automatic. It may seem superfluous to put the adjective 'subjective' before experience, because all experience is subjective, but this absence in schizophrenia provides a stark illumination of the phenomenon.

Hannah was a patient of mine with a diagnosis of schizophrenia whom I treated many years ago and who felt as if her internal experiences belonged to someone else. Her story has stayed with me for reasons that will become obvious.

Hannah

Hannah was a young woman who had begun to behave in an eccentric way some years prior to her admission. She moved away from her friends and family to fringe cultures, and gradually moved from this to reclusivity.

She had come to believe, and was firmly convinced when I began treating her, that she was experiencing the pain and suffering of a child who was locked in a basement. She had strange and unpleasant inner feeling states, sometimes physically painful, sometimes a choking sensation and sometimes distressing emotions that she came to understand as belonging to a girl child who was imprisoned. She believed that she was telepathically feeling the physical and emotional pain of the child, and she went to the police on many occasions to tell them about the girl. The allegations were investigated more than once, and no evidence was found to corroborate her story. One day she spotted one of her neighbours across the road and had an instant realization that he was the pedophile torturer.

A sudden realization like this, similar to the one experienced by Edith (in Chapter 1) on seeing the small gravestone and realizing that her baby was buried there, is characteristic of experiences in psychosis. We call this a 'delusional perception': a person sees something as it is, be it a gravestone or a neighbour, and simultaneously experiences a sudden transformative realization that is delusional . . . the gravestone marked the baby's grave, the neighbour was the evil pedophile. Back to Hannah's story:

Hannah told the neighbours and the police about the identity of the 'pedophile'. At this point her distress was uncontainable and her family and community were no longer able to cope; she was involuntarily detained to be treated in our hospital. On admission, Hannah insisted that what she was experiencing – the voices, the emotional distress, the internal sensations – were the experiences of the child, and she was deeply angry and frustrated that no one would believe her. She refused medication because it would take away her telepathic experiences and leave the child completely cut off. Hannah had been involuntarily detained with the same delusional system a few years previously. On her previous admission, the antipsychotics, from our point of view, had worked and had put an end to her abnormal sensory experiences. Following her discharge home from this admission, she stopped taking the medication, believing that it had removed her psychic ability to communicate with the child. As the medication left her system, the child, as she thought, began to communicate with her again. Hannah now believed that she had a 'sixth sense' that the antipsychotics suppressed.

There were endless negotiations with Hannah during the first few hours of this current admission about taking the medication, but to no avail. The child needed her, and the pedophile needed to be exposed – taking the medication would put the child in danger. We decided that it would be in Hannah's best interest to administer her medication involuntarily and keep her in the hospital for a prolonged period so that she would have a better chance to recover and re-engage in a world without a delusional system. We were in the ward office discussing her care plan, starting with the immediate intramuscular administration of an antipsychotic, when a nurse, an ashen faced young man, rushed in to tell us that Hannah had been cut down from the curtain rail around her bed. Hannah had tricked the staff into thinking that she was in the bathroom, and while the nurse waited for her outside the bathroom door she had returned to the ward and had hanged herself. She was dead when cut down. We found a note on her bed written in a quick scribble: 'Now do you believe me, help her.'

As the months passed, I reconstructed Hannah's subjective experiences in my imagination many times. She died, as she saw it, trying to save the child. How could she forget the years of the powerful sensate experience that had been transmitted to her from the child – it had been her every waking moment, the focus of all her attention, and the fodder of her memories. Treatment and a possible cure would lead to her abandoning the girl's one line of communication. Would her life be worth living if she abandoned the prisoner child? Could she live with that? She may have thought that there was no way to make the world believe her except through suicide. She may also have had a ghastly suspicion, on this her second involuntary admission, that she had a severe and enduring mental illness. If that was the case, she would have felt, as patients have said to me following recovery from psychotic episodes, that she had lost her mind. Either possibility may have been unbearable. Hannah again demonstrates that antipsychotic medication is indeed effective in controlling abnormal exteroceptive and interoceptive sensations, but it does not erase one's world view laid down by chronic psychotic experiences; something that I didn't appreciate the importance of at that time. Antipsychotics can control psychosis, but not memory. Hannah also shows how personal experience becomes a person's only filter through which they can understand the present world.

Hannah attributed her interoceptive feeling states to someone else, the phantom girl, as if they belonged to the girl and not to Hannah herself. In psychiatry we call these experiences 'made emotions', or 'alien emotions'. As we have already established in Part 1, we interpret our emotions through the hidden cortex of the insula in a vague body-in-insula map. A group of scientists in Toronto were testing pain sensation in patients who were conscious and about to undergo neurosurgery for epilepsy when they discovered something extraordinary about how humans feel pain.[88] Pain is often used to examine emotional/feeling circuitry in the brain because it is a relatively simple feeling to recognize and therefore to measure in experiments. The Toronto group did single-cell recordings in patients who were about to have a part of the big neural pathway that connects everything else to the prefrontal brain removed neurosurgically. This pathway is called the cingula: it is a big 'sweep up' circuit that connects the hippocampus, the amygdala, the insula and the cortices to the prefrontal cortex to be assimilated into a coherent whole.

The Toronto group applied mildly painful stimuli to the patients' skin and found that pain sensations predictably fired neurons in the insula. This was to be expected. What astonished them was that the 'pain neurons' also ignited when the patient watched the pinprick being applied to the examiner's fingers. Neuroimaging studies since have determined that there is a pain pathway within the brain going from the insula, where pain experienced in onself is mapped, and this circuits to the cingula system where pain for others is mapped, before being transferred to the prefrontal cortex.[89] [ii] Pain neurons represented one's own pain *and* the same pain in others. Because the neural system of recognition of emotions in oneself is the same one that is used in the recognition of feelings in others, it is referred to as the 'emotional mirror system'.[90] We literally do *feel for* others, and this comes from our feelings for ourselves.[91] The insula–cingula mirror emotional circuit is more active in individuals who are empathic and, as you may be predicting, is less active in those who have difficulty feeling for others, those who have anti-social or psychopathic personalities.[92, 93] We often blame psychopaths for their behaviour, but it would be more productive to teach them to learn what they are not wired to experience. The key role played by the insula in understanding the feelings of others fits with the evolutionary development of the insula, which increases in size and complexity from robins to dogs to humans, as emotional awareness and the ability to feel for others develops.[94] Even among mammals there is substantial variability in the anatomy and organization of the insula.[95]

The networks involved in mirror emotional systems are sometimes referred to as the social brain. I'm not keen on the parcellation of brain functions into 'social brain', 'emotional brain', 'thinking brain', and so on, because of the interconnectivity of the circuitry underlying experiences such as emotions, memory and cognition. These experiences are not separable in terms of brain circuitry but form a dense network that delivers a unified experience. Nevertheless, it is a convention that may help one to understand the fact that key constituent circuits within the whole brain network can be invoked to explain problems with specific functions, in this case emotional misattribution. The strongest evidence for involvement of the insula-prefrontal circuitry in mediating self and social awareness is in the brain disease called frontotemporal dementia.[96] Here there is a dramatic decline in self-awareness and social skills that is accompanied by a fairly specific and commensurate atrophy of the insula-cingula neural path.

Compared to schizophrenia, there is more general awareness of, and sympathy for, the difficulties in understanding feeling states that occur in autistic spectrum disorders (ASD). ASD is a very broad diagnosis, and many clinicians, including me, think that it has become overinclusive, extending to individuals with these personality traits who do not have a disorder. In any case, impaired socio-emotional awareness is a key element in real ASD, with difficulties recognizing one's own feelings and reading the feelings and intentions of others. There is evidence for reduced activity in insula-cingula areas in those with this diagnosis, as would be expected given the importance of this circuit in interpreting emotions in oneself and others.[97] It needs to be emphasized that something fundamentally different from the poor emotional awareness seen in ASD happens in schizophrenia. In ASD there is an overall impairment of socio-emotional awareness, while there is a total mixing up of one's own and others interoceptive experiences in schizophrenia.

The tangle of made emotions experienced by a psychotic individual should not be confused with empathy for others, which is an almost invariable human experience. The misattribution to others of feelings originating in oneself in schizophrenia, as experienced by Hannah, suggests that there may be problems with the emotional mirror circuitry in schizophrenia. Ray Dolan, a neuroscience psychiatrist from University College London, hypothesized about twenty years ago that there may be pathology in this circuitry in those with schizophrenia.[98] Since then, research has confirmed his idea, and there is now some more specific evidence that neuron function in the pathway from the insula to the prefrontal cortex may be disrupted.[99, 100] The finding that a specific familial gene in some people with schizophrenia is associated with reductions in this pathway supports this idea.[80, 88] This hypothesis, like most neuro-mechanisms of psychiatric disorders, remains speculative and is likely to be only one of many mechanisms that cause the psychotic experiences in schizophrenia. Made feelings, like those that Hannah had, form part of a pattern in a more general experiential problem in those who suffer from the acute psychosis of schizophrenia, where not only feelings, but also *thoughts* and *actions*, can be experienced as being implanted by other people or forces, and not originating in oneself.

Through a Looking-Glass

Through the Looking-Glass (1871) is a fantastic description of confused self–other systems. Whether Lewis Carroll was writing from his experiences after taking a consciousness-altering drug, or from unprovoked psychotic experiences, or from a brilliant introspective exploration of the feverish dawnings about consciousness in the late nineteenth century, has been the subject of much speculation. It is difficult to see how Carroll could have produced such an authentic description of psychotic experiences without having experienced them himself in some way. Alice seems to be subject to the agency of others as she is tossed around the imaginary chessboard, sometimes moving as a pawn, at other times being carried zigzag by the knight's movements, being prevented from expressing herself coherently by the queens, and so on. She worries that she may only exist as an imaginary figure in the Red King's dream and that she will cease to exist when he wakes up. She exists as a sort of reflection of others, like Hannah was for the imprisoned child.

What would happen next to Alice? Would the knight lift her up and gallop away, forwards or backwards, and then sideways to the right or left? Would the Queen be so vexatious, stealing her thoughts and replacing them with her own, that Alice will throttle her? Or has she been directed by the King to throttle her? It is easy to see how a person can become paranoid and angry with others in such states. Alice's experience of her actions being controlled – in her case by the anthropomorphized chess pieces – is called 'made actions'. Lewis Carroll's choice of a looking-glass as a metaphor for systems of made thoughts and actions was brilliant. He wrote the book

100 years before the first clue to the neural pathways of self–other discrimination came in the form of mirror neurons, although it was to be some time before the impact of this discovery was understood. The seminal 1992 study that led to the understanding that other people's actions may be represented in one's own brain came from a group led by Giacomo Rizzolatti in Parma, Italy.[101]

Their experiment, conducted on monkeys, looked at electronic recordings over the motor cortex part of their brains as they moved their hand muscles to form a grasp. The motor cortex is in front of the sensory cortex on the surface of the brain and is similar in design to the sensory cortex, the body being represented as a homunculus. The firing of specific cells in the motor cortex corresponded to the movement of matched hand muscles and gave the scientists a moving map of the motor cortex of Monkey's hand movements. During the experiment they observed something unexpected – as well as the motor neurons firing in a predictable pattern in the motor cortex when Monkey grasped an object, specific neurons in the *premotor* cortex, in front of the motor cortex, also fired. These premotor 'grasp' neurons were firing at the same time as their matched neurons in the motor cortex. What was the function of these neurons? They then discovered that, extraordinarily, if Monkey watched the experimenter grasping the object in a similar way to them, the same premotor grasp neurons fired, without the motor neurons. The pre-motor neurons seemed to *represent* the motor movement without Monkey actually moving, essentially allowing the brain to imagine a movement. The Parma group called these neurons 'mirror neurons' because they mirrored the actions of matched motor neurons involved in the grasp, the imagined grasp movement in Monkey and the movement in others.

The motor mirror neurons 'represent' one's own and other people's motor function, and also form part of a system in which one can distinguish self from other movement. Dysfunction in self–other distinction would make it difficult to know whether you yourself were doing something or whether someone else was doing it to you. It is subjectively terrifying, because the person so affected will experience their actions as being out of their own control, similar to Alice experiencing her movements as being at the mercy of anthropomorphized chess pieces.

Mirroring, Memory and Prediction

There is a further layer to the mirror motor neuron story related to memory. Mirroring is learned, and this memory is retrieved if stimulated by similar current experience. The motor mirror neurons fire not just in response to actual movements in oneself and others, but to predicted or imagined movements and feelings.[102] I think of mirror motor neurons firing off when I watch goalkeepers moving between the goalposts. They are highly trained to project from observation of motor movements to prediction of motor intention. Will the small movement of the striker's foot, thigh or eyes predict a kick to the left or right, high or low . . . or is the striker faking a movement to send the goalkeeper in the wrong direction? In a penalty kick, both the striker and the goalkeeper are playing with each other's mirror motor systems, predicting in the here and now, through intense observation, and in the abstracted representation of predicted movements, the others' mirror predicting systems. It is a dizzying mirroring of mirrored predictions. It is now known that the motor mirror system allows athletes to improve their performance through mental rehearsal – imagining movements leads to improved motor performance! The same prediction applies to emotional mirroring.

Adolescents *learn* that the emotions they experience are also experienced by others; this knowledge is not innate. The emotional mirroring system is brought to the prefrontal cortex and becomes a part of integrated, working memory systems. The Toronto scientists observed that if pain was applied more than once to the preoperative patients with the exposed cingulae, then the pain neuron in the cingula response preceded contact of the pin with the skin – they lit up in prediction of a sensation of pain. This is why we recoil at the anticipation of pain in ourselves and squirm at the idea of pain in others.

'Made Thoughts'

While through the looking-glass, Alice also had the experience that the Queen was stealing her thoughts and replacing them with her own. This experience happens frequently in schizophrenia and is called 'thought withdrawal' – the stealing of Alice's thoughts – and 'thought insertion' – the replacement of Alice's thoughts with the Queen's. There is a misattribution of the thoughts arising in one's own brain to another person or agency. I once treated a young man, Eoin, who believed that his thoughts were implanted by someone or something. He was never sure who the persecutors were, although he usually attributed the thoughts to anyone who was nearby. The thoughts were very unpleasant and were themed around sexual perversions. If someone was looking at him, he had the experience that they could read these thoughts. Eoin used to avoid making eye contact with anyone because, by keeping his eyes averted, he was protecting himself from others gaining access to the strange, disturbing thoughts. Using the metaphor of Alice's world on the other side of the looking-glass, Eoin not only thought that the King or Queen could implant thoughts in his head, but also that the thoughts so implanted could be accessed by all and sundry if they made eye contact. His mother told me that he would become agitated and perplexed when he looked in the mirror and would shout at and attack his own image. I asked him about this when he was better, and he told me that it was horrible to look at his mirrored image because he didn't feel that it was he himself. As he looked at his reflection, he was simultaneously experiencing auditory hallucinations – hearing out loud the implanted thoughts saying shameful derogatory things. Eoin lived in this confusion of distressing 'alien thoughts' – as it is sometimes called in psychiatry – during the first years of his psychosis, refusing help. It was only when he attacked a stranger at a bus stop that he was detained and treated in a hospital. It took him years before he could make eye contact with others.

Having particular types of auditory hallucinations – hearing voices speaking out loud, usually about you as if you are not there, or at you – is the most common pathological experience in schizophrenia. There is an interesting idea that the voices in psychosis may be caused by the misattribution of inner speech as coming from an external source.[85] If inner speech is a new idea to you – and I had some intuitive difficulty with the concept years ago, because it is not a prominent part of how I experience the world; some of us are more 'feeling' and/or visual than 'thinking' – just ponder what you are doing now when reading this book. You are, as you read this sentence, using inner speech to understand. It is estimated that 25 per cent of average waking time is spent engaged with inner speech. Speech and language are so complex, and sometimes so limiting, that even a brief exploration of the topic is beyond the scope of this book, so I'll not go there, except to say that inner speech is a representation of thinking processes. The theory of auditory hallucinations being the person's own inner speech attributed to some outside agency again points to the confusion about what is and what is not originating in oneself in psychotic states.

I have artificially broken down the integrated experience of awareness of oneself and of others into different domains – awareness of one's own and others' emotions through the emotional mirror system, awareness of one's own and others' movements through the motor mirror system, and awareness of one's own thinking as being a subjective experience. It is the integrated experience of these systems together that forms a sense of oneself – one's sense of self – in the world in the present and in memory. The relevance of this to biographical memory is that a biography can only start when an infant can experience awareness of themself. Self has to be recognized before it can be recorded as an entity. What we can learn from those with experiences of misattribution in schizophrenia is that the neural processes that are used to represent one's own movements, feelings and thoughts are the same as those used to identify these experiences in others, and that self- and other-experience need to be distinguished in order to have coherent experience and memory. Somehow there is an inbuilt neural mechanism that codes 'self' – as distinct from 'other' – experiences, and this neural mechanism seems to be broken in schizophrenia.

Of all the psychotic experiences, the most difficult to understand is the belief that one's subjective experiences – one's feelings, thoughts and actions – originate in someone else. You know that you, and not another force, is turning the pages of this book – you have motor self-consciousness. You can see or hear the words, knowing that the words have been written by the author – you have sensory self-consciousness. You do not believe that someone can put their thoughts as internal dialogue into your brain, or that you can put your thoughts into someone else's brain, or that everyone can access your private thoughts, sometimes to the extent that the thoughts are spoken out loud. You may have good emotional intuition, but you do not believe that your feelings are actually implanted by someone or something else. Although I have tried many times to vividly imagine experiences such as Hannah's and Eoin's, I find myself unable to do so. I have done the next best thing, I hope, and become familiar with this pattern of disordered experience and how it reveals itself to the observer. I can usually now quickly glean from what I observe that these experiences are present or not. Like all psychiatrists, as a patient once said to me, I am like a detective.

An Anarchy of Experience

One of the best descriptions of states of misattribution and confused boundaries comes from the psychiatrist R. D. Laing. In his famous book *The Divided Self* (1960), he articulated how the language of conventional psychiatry focused on *symptoms*, from the clinician's objective viewpoint, rather than the *experiences* of the psychotic patient. This wooden objectification of psychotic experience is something that is still evident in the discipline of psychiatry today, but I think that we are gradually moving to subjective experience as being the primary clinical focus. Laing's move from a symptom-based approach to an experiential one was radical, and he described for the first time the lived experiences of psychosis. 'It is often difficult for a person with a sense of his integral selfhood and personal identity . . . to transpose himself into the world of an individual whose experiences may be utterly lacking in . . . self-validating certainties.' One of the most basic of those 'self-validating certainties' is knowing that you are you, and therefore that *you* are.

Laing's insights into the experiential aspects of psychosis have been overshadowed in his legacy by his misguided beliefs about schizophrenia and his reckless treatment of patients. He became a *cause célèbre* because he revolted against conventional psychiatry, leading what later came to be called the anti-psychiatry movement. He believed that psychosis was a journey of personal exploration that could lead to a resolution with an improved sense of self. The belief that schizophrenia was caused by a past trauma was fashionable during the 1960s and the 1970s. If you could 'work through' the psychosis, it was reasoned, then you could find a cure for the psychosis. This was similar to Freud's belief that a range of psychiatric disorders, such as depression, anxiety and OCD, which he collectively called the 'neuroses', were caused by some past trauma and that, once the trauma was unveiled, then the disorder would disappear. To this end, Laing opened up a therapeutic house in central London in 1965 for those who had psychosis. Residents were taken off their antipsychotic medicines. There were no rules or personal boundaries, and patients, staff and visitors (often famous and always fashionably anti-psychiatry) frequently partied together. Patients took LSD and other psychedelics to enable what was seen as a journey of enlightenment from psychosis to sanity. In the ensuing chaos, which Laing himself called 'an anarchy of experience', two patients died by throwing themselves off the roof. Following these suicides the centre was closed.

The tragic end of his therapeutic experiment was entirely predictable. What is actually needed in those who are psychotic is a reinforcement of their own boundaries, a construction of their absent 'self-validating certainties', rather than further disintegration of the broken self–other boundaries caused by the psychosis. It is worth knowing that there are still a few psychiatrists, and many shaman-like self-proclaimed healers, who are opposed to treating schizophrenia with antipsychotics in spite of all the scientific and empirical evidence to the contrary. This oppositional stance occurs in all branches of medicine – oncology, for example – but it is more prevalent in psychiatry because of cultural bias and also the difficulties that a person has when psychotic of seeing themselves as being psychotic. None of us can be outside of our own brain; we can only try to make sense of the sensory information that is delivered to the sensory cortices, and thence to the higher brain for integration.

In hindsight it is now apparent that Laing was able to describe the fractured world of psychosis because he was a brilliant observer of his own experiences of psychotic disintegration. He experienced the melted boundaries that he wrote about possibly through substance abuse, notably the psychedelic drugs that the counterculture movement of the 1960s endorsed, and alcohol. In any case, his experiment in London, like those of the shamanic therapists, only heaped further and unnecessary suffering, sometimes unbearable, on those already lost in psychosis.

Mirroring and Memory

In the absence of functioning mirror systems, there can be no coherence in putting together input from the world and there is no coherence in the story that follows. The brain will continue to make dendrites and connect up neurons even if there is no resulting real-life *meaning*, because this is how neural mechanisms operate. The brain does not stop organizing because the input will have no rational overall meaning. Memory will continue to be formed because cell assemblies will be put together in the hippocampus and will be transported to the cortex in the dynamic functional connectivity that is the biological life of the brain. Usually, the person with chronic psychotic experiences will, over time, form some sort of narrative to explain their experiences. Hannah's narrative to explain her alien emotional experiences was that it was a telepathic implant from a captured child. Sometimes this putting together of experiences to make a narrative has some superficial plausibility, but this usually falls away quickly on closer scrutiny. Whether psychotic or not, subjective experience becomes biologically embedded in the messy web of neural connections that is memory. The basic principle that current experience gets laid on to memory networks through association applies as much to psychotic as to non-psychotic experience, and as time goes by the psychotic individual's memory and way of being in the world will become increasingly estranged from common reality. As the delusional web of experience becomes elaborated, the more difficult it will become to bring them back to the shared reality of normal experience.

In this chapter we have followed the origins of biographical memory from the amnesic infant to the dawnings of self-recognition, and of how self-recognition has evolved phylogenetically and is reflected in infant neurodevelopment. Self-recognition is just the start of consciousness of 'self' and the beginning of the separation of self that continues to develop throughout life. How the infant with inklings of self-recognition will eventually develop a sophisticated system of self-awareness and a sense of others possessing that same quality is not yet clear, but it happens in concert with the increasing complexity of memory organization. We will now look at this developing complexity.

10. The Tree of Life: Arborizations and Prunings

To exist is to change, to change is to mature, to mature is to go on creating oneself endlessly.

Henri Bergson (1859–1941)

A person's way of being in the world changes as they age, a process that starts at birth. This is because the brain develops an increasingly complex way of understanding the world – the outer world and one's inner interoceptive world. Developments in the organization of memory systems can be seen in changing patterns of subjective perception that mark the different stages of life. The general life-cycle pattern of human memory is reflected in the changing patterns in overall brain anatomy. The structural and functional changes in brain systems continue throughout life, but there has been more emphasis on early brain changes because these are foundational and generally have a reliable trajectory into adult life. Let us start by looking at a *relatively* simple development in brain memory systems, what I call the Beethoven phenomenon.

The Beethoven Phenomenon

Beethoven (1770–1827) began to lose his hearing during his twenties but continued to compose music until he died. Some of those compositions considered to be his greatest were created during the period of his most profound deafness.[i] How did he write music when he could hardly hear sounds? The answer lies in the location of his deafness, which was found at autopsy to be in the auditory nerve – the sensory nerves conducting sound from the outside world to the auditory cortex of the brain. Beethoven's auditory nerve was diseased, but his auditory cortex, which had memorized sound, notes, tunes, music, had remained intact.

When listening to music, the auditory cortex lights up. When we *imagine* music, in addition to the auditory cortex the prefrontal cortex also lights up.[103] Imagined tunes are sound memories projected from the prefrontal to the auditory cortex. Following the gradual loss of sound from the outer world, Beethoven was hearing through neural representations. The auditory cortex–prefrontal dance was the composer of the virtual music. Beethoven provides an amazing example of auditory sensory experience being transformed into cortical memory systems, and of how the cortical memory can then be played out in the prefrontal cortex as a sort of 'inner' sensory experience. His creative genius derived from some form of staggeringly complex cortical representation of musical notes and their configuration into beautiful sound patterns.

The Beethoven phenomenon is a highly developed example of a universally experienced phenomenon. In his rigorous musical training from a very young age, Beethoven had created a neuronal labyrinth of musicality in his auditory cortex. The process of gathering and organizing sensory memory takes place at a galloping pace in the expanding cortical networks during childhood. Children are immersed in the sensory world, relatively free of the abstract 'short cuts' that develop with the more organized cortex. The absence of abstract thinking is apparent in the way that children tell a story. Events follow events without being assigned meaning . . . they have memorized images in chronological order and they recall them in the same sequence. Their output of non-assimilated imagery, the absence of context and meaning, is delightfully naïve and sometimes illuminating, because it gives us an insight into our learned ways of looking at the world.

Because of this, the world of the child is constantly new and more immediately sensate than it is for adults. Dylan Thomas's dazzling description of the sensate experience of childhood in his poem 'Fern Hill' needs to be heard in his sonorous Welsh voice to be fully appreciated. You can feel the vivid sensate excitement of childhood through the melodic images. It is all movement.

> All the sun long it was running, it was lovely, the hay
> Fields high as the house, the tunes from the chimneys, it was air
> And playing, lovely and watery
> And fire green as grass.

Every time I read this dancing romp of a poem I am back in my own Fern Hill – my mother's family farm in rural Kerry, where we spent our childhood summers. I can almost smell the cut hay, feel the itch of it, and see my Uncle Jim looking down like a giant with a huge hand around a jam jar of heavily sugared cold tea. What is happening in the neural circuits during the expansion of sensate learning during childhood?

We do know that there are changes, visible to the eye, in brain structures that reflect the sequential stages of development in cognition and behaviour. Cognitive and behavioural changes are more culturally familiar than memory systems, but it is the changes in memory organization that underlie the cognitive and behavioural changes. The developing infant is taught in ways that are appropriate to their stage of brain development. The infant, immersed in a sensory world of tactile, visual and auditory learning, where everything from their mother's nipple to their own feet are tested and tasted in their mouths, is taught simple sensory information – naming objects and people, colours and sounds. This is the period of rapid change in brain sensory systems, which can be seen in the expansion of the volume of the sensory cortical brain.

In a foundational study from UCLA in California and the National Institute of Mental Health in Maryland that tracked changes in the child's brain, researchers followed up thirteen children ranging in age from four to twenty-one years.[104] Cortical development from childhood to young adulthood was mapped every two years with sequential brain neuroimaging scans. While they found great variation among the children, there was also a common pattern of development that followed the evolutionary development of the brain from phylogenetically older to younger animals. The general pattern started with increases in the size of the sensory and motor cortices. This growth reflects the rapid expansion of relatively simple sensory and motor learning.

Neural connections are then elaborated among the different sensory cortices, allowing multisensory perception to develop so that sound, sight, touch, taste and smell can become integrated. A child will eventually be able, on hearing a bark, to turn their head in the direction of the sound, expecting to see a dog. Visual depth perception, 3-D vision, is a good example of the organization of multiple cortical areas, including visual and directional intelligence. Individuals without sight from birth cannot imagine in 3-D because they do not get the necessary input from the seeing world to elaborate 3-D perception. As in the history of visual art, as with individual neurodevelopment, perspective is learned.

The second important aspect of brain development is the process of 'pruning' of the dendritic connections among the neurons. It is called 'pruning' because it is a similar process to cutting back a fruit tree to maximize fruit production. Neurons are pruned to maximize precise output.[61] You might have thought that the dendrites expand as knowledge expands, but, counterintuitively, we are born with too many dendrites and continue this process of overproducing dendrites during the first year of post-birth life. At a cellular level, the tiny baby's neurons are overconnected, and they are susceptible to being overloaded easily by too much sensory input, because their neurons are firing off in all directions in a haphazard way. So we soothe, and we teach simple sensory information; we direct and contain. Sensory neurons are the first to be pruned following this brief expansion. This is reflected in the experience of sensory learning, which is really a process of discrimination. For example, a child will come to recognize a tree as a tree, and not just another plant, and then will learn whether it is, for example, a big or a small tree. Later they may learn that some trees lose their leaves in winter, and others are evergreens. They may come to know their trees and identify the genus from the leaves, the size, the way the branches grow, the bark, the flower or fruit or a combination thereof. At this point they have developed good cortical networks that allow patterns of recognition, and more importantly discrimination, among trees. They no longer need to scrutinize the constituent parts: the tree can be identified at a glance.

The real becomes represented, filed away in the cortex as an increasingly precise abstract representation that can then be automatically processed. These shortcuts are necessary in order to process more refined information, otherwise we would spend unnecessary time having to process all the pieces of information that had been learned throughout one's life each time we identified a tree. In this part of the world, most people would identify a holly tree at a glance. You may hear a rustling in a holly tree in winter and predict that it is a mistle thrush, nested there for winter because the berries provide food and the prickly leaves protection. I intuited that the rustling in my holly bush this winter was a mistle thrush because I saw a larger than average thrush, which I learned was a mistle thrush, in a holly tree in my garden during previous winters. One is carried by multi-sensory perceptions – the sound of a rustling, the movement of holly branches – to immediate perceptions based on prior knowledge. A dendritic connection has formed between a rustling in a holly tree and a thrush to form a perception, facilitating a connected-up experience of these sensations to a shortcut conclusion. One may look at a distant tree in Africa with white flowers, and be prepared to see, as I was, not white blossoms but a fluttering of white butterflies. The memory of a 'butterfly' tree had been laid down years previously, but I had not seen many trees in Africa with white blossoms. We expect to see a car when we hear the sound of a car, a truck when we hear the loud growl of a big diesel engine, a racing car when we hear that velvety growl. These automatic perceptions and predictions are the 'ways of seeing' that Berger described, alluded to in Chapter 3.

As an infant learns about the world, the connections among the neurons in the cortex become less dense. Pruning in the sensory cortices peaks at about three years of age but continues throughout childhood at a declining pace. The prefrontal brain develops at a later and slower pace, because it is the principal brain area where inputs from the cortical sensory areas, from the amygdala-insula and from the hippocampus, come together to be sorted out. It is where information from multiple areas is 'held' to be manipulated. It is the expert juggler that can, stealthily in silent mode, or in active working memory, transcend the external sensate world, to imagine or to predict, to create or to manipulate. The ratio of prefrontal brain to total brain volume is biggest in humans compared to all other species, and as human development tends to follow evolutionary development – the principle underpinning phylogenetics – this area undergoes substantial development during adolescence and early adulthood.

Prefrontal Pruning

The development in the prefrontal brain can be seen in changes in the prefrontal cortex that commence during late childhood, when the synapses are being pruned to form stable neural patterns for understanding the world.[105] Pruning cuts over-expansive spread of signals and creates neural pathways for discriminating between signals, allowing the development of systematic thinking, or abstract reasoning. Pruning facilitates the organization of vast amounts of input, enabling the developing brain to take shortcuts through learned pathways of knowledge. Prefrontal pruning is the neurodevelopmental signature of the adolescent brain, but it continues to occur throughout the twenties and thirties, something that has been discovered only relatively recently.

The second major change is a growth in the white matter that also takes place during brain development. The brain is made up of patterns of grey and of white matter: grey matter is composed of clusters of neurons, and white matter is composed of the axons that emerge from the neurons to carry the signal down to the dendrites and into the next neuron. The axons are white because of an important process called 'myelination', in which fatty spirals of myelin – which look like microscopic swiss rolls – are wound around neurons.

These fatty cells provide insulation that speeds up the signal transmission, up to 100 times that of an unmyelinated neuron. Myelination halts the promiscuous firing of neurons with each other as the signals pass down the axons and is part of the process of targeting signals to go in key directions. What previously was a relatively dissipated signal becomes directed because, firstly, some synapses are reinforced and others wither, and, secondly, improved neural insulation speeds up the signal transmission.[ii]

The development of discrete pathways through pruning and neural insulation means that some connections between neurons are sacrificed at the cost of reinforcing others. The general scheme during early development is that neurons arborize in a promiscuous way during prenatal and early postnatal brain development to maximize foundational information, and then become pruned to discriminate and sharpen sensory input during childhood. In later childhood the prefrontal dendritic shoots get pruned to provide cognitive and emotional ways of understanding the world. Some interneural pathways are strengthened because of being frequently fired, while others fade, leading to *relatively* fixed and automatic networks of interpretation.[106]

The pattern of initial over-elaboration of new information occurs with all new information and throughout life, and it will be familiar to anyone who has tackled a new area of knowledge, and not just lifted information from the internet. There is an initial period of being lost and overgeneralizing in the sea of novel information, before emerging with an organized, contextual understanding of the new subject: knowledge expansion followed by knowledge discrimination.

Neurodevelopmental Disorders

Cognitive brain development in the prefrontal circuitry that occurs during adolescence and early adulthood may not proceed as it should, and there is a focused interest in recent years in how pathology in pruning processes may lead to developmental disorders, such as ASD and schizophrenia. The experiences and behaviours typical of schizophrenia generally emerge during adolescence and early adulthood. An adolescent does not wake up with schizophrenia – it develops over years. As might be predicted, if brain processes do not get organized in the normal way, working memory will be disorganized. Some researchers have hypothesized that there may be excessive pruning in schizophrenia, and reduced pruning in ASD.[107] A group working in Cambridge have shown that the genes involved in pruning and myelination are known to be risk genes for schizophrenia.[108] While highly speculative, these lines of inquiry have evolved into what is called 'network pathology', as distinct from the single pathway pathology in dopamine neurotransmission that was naively believed to give rise to schizophrenia, which we will examine later. Rachel was a patient of mine who had onset of psychosis during childhood and did not have normal integration of sensory input. Her world was incoherent to others, and her thinking and speech had no continuity of meaning. She had minimal autobiographical memory.

Rachel

Rachel had been very psychotic since childhood and had, for her own safety, lived in a psychiatric institution since she was a child. In addition to severe treatment-resistant schizophrenia, she had bad epilepsy. She had been referred to me during a period in my career when I worked in a general hospital alongside neurologists. Her family wanted to trial her on clozapine, which still remains the best antipsychotic available. Clozapine can only be given when all other treatments fail, because it can cause serious side effects, one of which is to destabilize epilepsy. She was already on olympian doses of both antipsychotics and antiepileptics but remained very ill, and her family were desperate. We decided to admit her to the acute neurology ward, where her epilepsy and psychosis could be monitored continuously, to commence her on a trial of clozapine.

Rachel is one of the few people whom I have treated who held convers-ations with herself but not with others. She sat in a chair by the bed, or in the bed, talking to herself in different voices while the nursing staff chatted away to her as they brought her through the daily ward routine. We spoke to her every day and never received an acknowledgement that we had said anything, except for an occasional hostile tone in her speech and a glower-ing expression. We were of course rudely interrupting her hallucinatory conversations. She spoke in the voices of whoever she was on the day, sometimes having a deep male voice and sometimes a female voice. Her conversations made no sense, but an occasional glimpse of a connection with the world would peep through her ramblings. She told us once that she was Julius 'Seizure' – a pun referencing her epilepsy that made us smile and wonder whether an underlying innate character had formed in spite of her crippling psychosis. Sometimes she was Marie Antoinette or some other European royalty. On one occasion she told us that she was Grainne 'Whale'. This was no doubt a reference to the sixteenth-century Irish pirate Grainne Uaile (pronounced 'wail'), a legendary female in Irish mythology.

We were into a few weeks of treatment with clozapine when the scene that I remember, reinforced by my frequent recollections, happened. We walked into her room and she made eye contact with me for the first time, said good morning and smiled shyly. I remember holding myself very still. We were transfixed, hardly daring to believe what seemed to have happened. I conducted a short conversation with her about what she had eaten for breakfast . . . was she comfortable? . . . had she slept well? . . . did she like the ward? She responded appropriately, and we slid from the room when I felt that she may have had enough normal talk for day one.

I remember the team standing in the corridor of the neurology ward immediately afterwards looking at each other, not speaking, sharing a few emotional moments. In my early career I didn't allow much through emotionally, but that has changed. It may seem to an outsider that the drift would go in the opposite direction – from one of struggling to deal with one's subjective emotional responses to such suffering, to one of objectivity. What happened actually is that it took me about a year to cope with the emotional intensity of the work. That first year was exhausting, and then I learned, as clinicians do, to get on with it and do my best. Now I find myself, after thirty-seven years in the field of clinical psychiatry, at unpredictable moments being hit viscerally in a way that I used not to be. We may bow down in respect at the sheer human endurance that we see in the very ill, but to suffer in a world that is hostile to this suffering . . . that is what hits the punch now.

Rachel went on to emerge with a charming, child-like quality from the psychosis. Her coming into herself did not appear complex, and she made no reference to fictional identities once the psychosis resolved. She had little autobiographical memory and only spoke in a vague way about having been in the hospital. Although her event memory was very sparse, she had a good vocabulary, was literate, seemed smart and recognized everyone. Shortly after her psychosis resolved, Rachel was discharged back to her local psychiatric hospital and we didn't get a chance to observe her full emergence from the shut-off chaos of her lost childhood to the brain organization that was, somehow or other, released by clozapine treatment. Her mother wrote to me until I left that consultant post. Rachel continued to improve, and a few months after discharge from us to her local hospital she returned to live with her family.

Rachel's brain was scrambled by the confusions of psychotic sensation, and/or by disorganized integration of sensory signals. Her memory networks were not being organized in a way that could lead to a coherent processing of the neural input from the world, or to a coherent narrative, to any narrative. The absence of normal organization of memory in Rachel demonstrates how the developmental organization of brain networks, absent in her, is the foundation for biographical and working memory. Rachel did not really know who she was until she was able to record experience in a way that made sense.

Although we do not understand how, her response to clozapine liberated her dramatically from the chronic and crippling experiences of thought disorder and hallucinations. When asked how antipsychotics work, the inquirer frequently follows with, 'Has it something to do with dopamine?' Dopamine is probably familiar to most people as the 'reward' neurotransmitter. At one point in popular culture, dopamine release was being invoked for almost everything enjoyable and seemed to be a common neural pathway for all human pleasures. The dopamine omni-pleasure oversimplification is highly unlikely, if only because dopamine constitutes about 1 per cent of neurotransmitter content in the brain. Generalizations about neurotransmitter function may be attractive to explain experience and behaviour, but they are usually wrong.

The dopamine circuit does have a role in reward, because it is probably the final common pathway for many brain-reward processes, but there are a myriad of circuits that converge on the dopamine circuit, all of which could mediate the reward and any of which could be problematic. Dopamine does not cause the 'hit' – the chocolate pleasure, the heroin high, the orgasm, the alcohol buzz – but it is the neurotransmitter in the pathways that get laid down to memorize the hit. A common mechanism of action across all antipsychotics is the reduction of dopamine transmission, which explains why for decades the dominant theory in schizophrenia was that dopamine systems were overactive. Our thinking on how reduction in dopamine neurotransmission relates to the therapeutic effects of antipsychotics is very speculative as yet, but it can be conceptualized for ease of remembering as a process in which there is a filtering out of sensory information being transmitted through to the prefrontal cortex. The filtered information coming through to the integrative brain can then be put together in a more coherent way.

As neuroscience expands to an understanding of the brain as an indivisible connected-up mega-tangle of neurons, schizophrenia has now moved away from a simple explanation of overactive dopamine neurotransmission to that of disorganized network pathology that usually starts during brain development.[108] I was introduced into pioneering neurodevelopmental research in the 1990s by Robin Murray, a world leader in bringing ideas about neurodevelopment into the neuroscience of psychiatry and subsequently into mainstream psychiatry. The experiences and behaviours typical of schizophrenia generally emerge during adolescence and early adulthood. An adolescent does not wake up with schizophrenia – it develops over years. I was lucky enough to be a lecturer in the Department of Psychiatry during Robin's tenure as Head of the Institute of Psychiatry. He carried a small notebook in the breast pocket of his jacket in which he recorded the dates of birth of our patients. He was particularly interested if a patient had been born in 1958, because this was a year of a big influenza. He had hypothesized that the flu virus, or a protein from the mother's immune system, may have entered the developing brain of the fetus causing a neural mis-wiring, in particular in the prefrontal neural circuits that use the neurotransmitter dopamine. Prefrontal circuits using dopamine develop significantly during early adulthood, and consequently the brain problems, he reasoned, would not become really apparent until dopamine circuitry was being reorganized in adolescence, potentially leading to schizophrenia-like illnesses.[109]

The idea that a uterine virus may cause some forms of schizophrenia was considered far-fetched back then, but research since has confirmed that the prenatal environment is a foundational one in terms of neural wiring, and that schizophrenia is probably a neurodevelopmental disorder of brain wiring. An important point is that mis-wiring and misfiring in integrative circuitry can follow from multiple causes, including infection in the pregnant mother, brain injuries acquired during early development or childhood abuse or neglect, or substance abuse during childhood or adolescence. The effects of the immune system, inflammation and antibody formation on neural systems have since become mainstream subjects in neuroscience.[iii] Psychosis can also follow inevitably from an inherited genetic DNA sequence that codes for a faulty brain protein. Or, probably more likely, psychotic disorders may follow from a combination of multiple causes with small effect sizes that cumulatively fail to bring a developing brain to a critical level of coherent networking.[110, 111]

Abstraction and Imagination

Sometimes we are aware of our own layering of patterns of discriminatory learning, like learning about trees, and sometimes we are not aware of the knowledge upon which our sometimes correct, sometimes approximate, sometimes incorrect intuitions are derived. Henri Bergson believed, based on subjective observation and introspection, that intuition was grounded in memory, and he was correct. Intuitive knowledge may seem to be speculative, but it is not a wild guess – it is grounded in hidden information that has become automatic. You know that you know something but you may not be sure of how or why you know it. The immediate intuition that the rustling in a holly tree may be a mistle thrush is built on the layers of cortical association memory that are activated when I see and hear this combination. The intuitions that one has are the outputs from current brain inputs pitched against neural patterns in working memory.

The ability to reason, to use information in an abstract way, develops in tandem with prefrontal brain pruning. Information becomes more permanently organized into neural patterns as myelination progresses during early adulthood and relatively fixed ways of thinking, imagining and feeling, of generally being in the world, progress. With prefrontal myelination comes the ability to take shortcuts through abstract information.

In older adulthood, reasoning and prediction improve, and sensory function but not, importantly, sensory appreciation, declines. Little is known as yet about the brain changes that occur from middle to late adulthood, except that hippocampal volume and hippocampal efficiency decline, as part of the heave from the exteroceptive to an abstract inner world. An important caveat is that today this slide is far less likely to happen in early old age because of improved treatments for sensory deficits, such as routine cataract removal and more sophisticated hearing aids.

The Older Adult and Deep Knowledge

The changes in thinking patterns and ways-of-being-in the-world that occur in the transition from adulthood to older age reflect shifts in memory dynamics from the all-consuming sensory expansion of youth to the intuitions, sophisticated creativity and wisdom of older adults. The cortex continues to thin as adulthood proceeds.[112] The advantages of this shift do not as yet receive much attention, but it is likely that this will change as longevity extends. As we age, sensation is no longer knocking down the door to swarm the brain: it is being increasingly deftly processed without much attention being applied. Adults may even get to the point where they automatically process, and fail to appreciate, the over-familiar beauty in the natural world or the sensory vitality in a city lived in for decades. Yes, it would be nice, one may sometimes think, to return to the days of childhood, to the simplicity of experiencing, rather than interpreting, the living world. But we inevitably fall out of the relatively uncomplicated interpretations of childhood to the more layered interpretations of maturity. One could take a sentimental view of what is often called 'innocence' – the Romantic era in literature was steeped in the idealization of lost innocence – but the refining of prefrontal networks in adulthood brings with it an improved ability to understand and to predict, and overall greater self-efficacy and self-realization. This wisdom, if it develops, can bring great peace of mind, and stability for society. In folklore, as in life, the wise are usually old.

Pushing into older age, sensory systems go into decline. Memory as a neurophysiological process becomes less efficient. Older, compared to younger, people have poorer short-term memory but are better at problem solving and guessing. There is a trade off between abstract prefrontal facilities improving while the capacity to make memory deteriorates. This allows individuals, from early to later adulthood, to process information in different ways that are both successful. The perspectives, one derived from better working memory, and one derived from deep knowledge, are complementary to a functional society.

Observations from the natural world can be pared down in physics to universally applicable laws of gravity, matter, sound and motion, entropy and events, and so on. It takes unimaginable cleverness to pare down the entropy of the natural world to equations and principles, but *this* is the essential process that we, each of us, use to understand life as lived. There is a point where, in the putting together of information, a filigree of networks emerges from years of input and successive dendritic refinement. This is the point where deep knowledge is possible. Contrast this deep knowledge acquired through the layering and refinement of vast amounts of experience with a situation that commonly arises today, in which readily available information can apparently make one into an expert overnight.

If one lives to old old age, having not been felled by disease, some sort of final sensory and experiential slide from the world often happens. The person, progressively cut off from the world by the decline of sensory systems, seems to come to a point where there is a relinquishing of the sensate world. Thankfully this is being warded off more effectively and for longer as sensory aids improve. Fragility leads to reduced mobility, and there is a distancing from the momentum of the world. Sándor Márai in his novel *Embers** has described, with exquisite pathos, the final life slide from the sensate world to an abstract one:

> ... everything gradually becomes so real, we understand the significance of everything, everything repeats itself in a kind of troubling boredom ... That is old age ... Gradually we understand the world and then we die.

Perhaps this passage resonated with me because I had a hovering

* Sándor Márai, *Embers* (1942; Viking, 2001), pp. 193–4.

sense for many years, a subliminal dreaded anticipation as the years passed and I began to understand patterns in life, that I would switch from a sense of the possibilities in life to inevitable prediction, into 'a kind of troubling boredom'. And then one day, somehow or other, I was not anticipating the dreaded staleness but instead was appreciating the rich sensate world in the present. One returns to the world of sensation – not the headlong hurtle of youth, but a richly nuanced one that you want nothing from, except to be in it.

I will finish with a word about imagination and creativity. As Patrick Kavanagh wisely wrote, 'imagination is the blossom on the stem of memory', bringing us back to where we started, the Beethoven phenomenon. His astonishing gift of creating some of the most beautiful music ever composed when deaf resulted from a highly developed auditory memory interacting with a highly functioning prefrontal conductor. At the end of the day we are, even if imagining at the level of Beethoven, individual constructs of sensory-memory systems of infinite complexity. One may *feel* that that the experience of living is more than an unending shifting network of highly sophisticated, infinitely arborized neurons, that there is something beyond oneself, outside of memory and even imagination. The fact is that this feeling is created by neural networks that make you *feel* as if you are more than 'just your brain'. This is the ultimate feat of abstraction – the representation of oneself, a consciousness of oneself as an integrated extant being in a world of other similarly wired humans. In the chapter that follows we will explore these higher states of consciousness.

---

* Patrick Kavanagh, *Collected Prose* (MacGibbon & Key, 1967).

11. A Sense of Self

One of the concepts most fascinating to humans is that of 'higher consciousness'. We think of higher conscious experience as being the holy grail of human exclusivity. We tend to think of memory as something separate from consciousness, but, as Henri Bergson wrote, 'there is no consciousness without memory.' In any individual life, all going well, consciousness will evolve hand-in-hand with memory, and, as memory systems become more complex and integrated, so too will systems of awareness and consciousness. As we've seen, self-recognition is the start of autobiographical memory, so from the very start of life awareness of oneself is bound up with memory of oneself. Systems of awareness then develop from self-recognition to complex representations of oneself, usually referred to as 'higher consciousness'. Eventually, a person will become aware of their own consciousness: this is called 'meta-consciousness'. Meta-consciousness is, in essence, looking at yourself looking at yourself.

Science does not always provide the best language for understanding complex, integrative brain function, particularly in the domain of higher levels of consciousness. This is not helped by the scientific parcelling up of brain functions into memory, cognition, emotions and so on. It can be more instructive to look at what the arts have to say if we are to understand memory in the broader context of human experience. Indeed, in the historical journey towards an understanding of higher levels of conscious experience, in the last 150 years or so the sciences, through piecemeal discoveries, and the arts, through alterations in formal expression, have danced around the same theme of consciousness. The James brothers, who lived during the transition from the nineteenth to the twentieth centuries, are a rare and beautiful example of an intimate dialogue between the arts and the sciences in the area of memory and consciousness. The great novelist Henry was among the first to move from event narration to 'narrative consciousness', where the narration comes directly from the perspective of the character's inner experiences.[i] His brilliant psychologist brother William moved the psychological concept of memory from one of flat knowledge to one of live streaming the dynamic inner world of conscious experience. It was William the psychologist, and not his novelist brother, who coined the phrase 'stream of consciousness' that was to become a key phrase for the modernist literature of the twentieth century, culminating in the famous stream-of-consciousness portrayal of a day in the life of Leopold Bloom in James Joyce's *Ulysses*. What is a stream of consciousness if not a record of present interospective experience understood in the context of past experience?

Psychiatrists were also navigating the new waters of consciousness during the *fin de siècle* of the nineteenth century. Although Freud is seen as the giant of psychiatry in expanding our understanding of consciousness, his concept of consciousness was generally limited to *past* memory, and he failed to incorporate the ongoing vital momentum of present experience into concepts of consciousness. His theories of conscious and unconscious memory, of repressed memory, lack the immediacy of the rush of present experience that marked the intellectual excitement of the James brothers' writings or later existential literature. Freudian consciousness was more a structure than a dynamic form: he wrote about consciousness as if it were layered, from the depths of the unknown unconscious through the subconscious to consciousness, like a bathymetric map in which darker blue shades indicate greater depths. The immediacy of consciousness in which experience moves from one moment to the next, the stream of time, is what the Jameses explored with great flair, and what Freud missed.

The movement of consciousness is easily overlooked because of the obvious fact that it is only in the *present* that consciousness exists. We have explored the theme of the present in Chapter 7, and concluded that the past and the future only exist in memory, while what we call the present is really consciousness. Higher consciousness is a living process, in which there is an exchange between live-streaming systems of sensory input and memory networks.

Abnormal States of Consciousness

Expanded states of consciousness provide a unique insight into normal states of consciousness. Even before the James brothers, Dostoevsky (1821–81) was writing about consciousness, particularly abnormal states of consciousness. For me, Dostoevsky brings a sense of amplified consciousness, a rushing immediacy, an uncomfortable excitability, a profusion of sensations that is like a manic state. He wrote, in *Crime and Punishment*, '. . . to be overly conscious is a sickness, a real thorough sickness.' When I first read Dostoevsky, I was aware that his descriptions of subjective states were distorted, but I, like countless readers, found this compelling, in the way a horror movie might be. How could one describe this stream of distorted consciousness so convincingly without having had the experience of abnormal states? And Dostoevsky did in fact experience abnormal mental states: he suffered from epilepsy, in which psychotic experiences frequently occur because bits of the brain are randomly firing off.[113] Before his seizures he sometimes had a few moments of ecstatic, heightened consciousness, in which he felt that he transcended time. He describes such states in many of his novels – for example in *The Idiot*, in which he writes of Myshkin, 'His sensation of being alive and his awareness increased tenfold.'

Arav was a patient of mine who experienced this 'real thorough sickness' in an episode of mania. He has a Dostoevsky-like gift for vividly describing, orally and in writing, his subjective experiences.

Arav was twenty-one years old when he was brought by the police to our hospital. He had been depressed for what seemed to him most of his life. When he became manic, some weeks before his admission, he 'thought that the psychosis was a cure for the depression,' as he told me when he was admitted. It had been for him an overwhelming experience: '. . . the best time in my life, and I wanted to communicate this to everyone. I was in the middle of a big revelatory experience, and if I didn't communicate it, it would be as if it hadn't happened.' He began talking in an uninterruptable way about what was happening to him. He felt that barriers to others were melting away, that he could communicate telepathically and that he was 'filled with power from other people'. Everything from the world was 'so fascinating and I was in hyper-space'. He said that it was very strange to be 'too aware', and that everything that was coming into his head had a special significance. He believed that he could change people and fix all the problems in the world . . . 'cure the dying, get scientists to figure out why people don't optimise each day . . .' He had a 'multitude of ideas that were always whispering in the back of my head and I was trying to find the rhythm'.

He followed people around trying to tell them about his hyper-perceptive experiences and his 'tangents of insight', so that they could share his new knowledge. He did not sleep for the week prior to presentation. In the police station he was examined by a doctor who directed him to our service. Arav found himself in the bizarre situation where he was in a euphoric state of heightened consciousness and the doctors were telling him that he was ill. He was told that he lacked insight, whereas he felt that he had 'too much insight'. He did accept medication because he saw that the experiences were getting out of control and that he needed some medical help to sleep and become less stimulated. He gradually, over some weeks, became less overactive and less overtalkative. Arav remained preoccupied with the powerful experience of his mania, and would talk about little else during his admission.

Shortly before his discharge from our inpatient unit to the day hospital, he told me . . . 'Why do we see colours and dogs see in black and white? We see because we build things up . . . it is all just sensory input.' He went on to explain that the heightened sensory input that he had during his psychosis was a gift that had got out of hand, but that he became 'so aware that he [I] was scared'.

Arav's wonderfully vivid descriptions of the hyper-conscious state of mania provide an example of heightened experiences of sensation, called 'hyperperception'. As well as the stream of heightened neural input coming from the outside (exteroceptive) world, there was another heightened stream of neural input coming from his body – the interoceptive stream. Arav told me that when manic he 'had a new consciousness' of his body. Along with exteroceptive hyperperception, he was experiencing intense interoceptive sensations being delivered from his body. Just to recap, interoceptive sensation is first brought to the insula to be 'mapped' in the body – the insula will identify pain and the location of pain. From the insula the neurons go to the prefrontal brain to integrate one's feeling states into working memory. The insula-prefrontal drive is Yeats' 'rag and bone shop of the heart', containing one's unique history, which can give voice to the most finely calibrated emotions, the muted pleasures and pains of one's remembered, or one's imagined, life. It is what is left when all the trappings of vanity and self-deception have evaporated in older age. Raw feeling states are introduced to consciousness, although this interoceptive awareness may be ignored. In mania there is an uncontrolled gushing of activity in this system, a highly sensitized self- and other-feeling system that leads to the experience of melted boundaries between oneself and the world, and an exaggerated feeling of connectedness. This stage of mania is usually short-lived and evolves into a state of frenetic, uncomfortable over-activity and over-arousal.

Psychedelics

Arav's 'real and thorough' sickness of consciousness was preceded by feelings of euphoria. People have sought out such euphoric states for thousands of years. An expanded sense of connectedness – with oneself, with other people and with the material world – is typically experienced when tripping on psychedelic drugs. The writer Aldous Huxley took the psychedelic mescaline in 1953, before psychedelics were controlled drugs, and subsequently wrote a book about his experiences called *The Doors of Perception*. It remains probably the most famous account of how psychedelics heighten self-awareness, perception and connectedness with the world. These experiences are commonly spoken about as being religious, spiritual or mystical. I want to emphasize that there is a fundamental difference between the expanded consciousness of psychedelic-induced mental states and the uncontrolled hyper-consciousness of mania, and I am only using both to exemplify heightened experiences of consciousness and not to make comparisons.[ii] It is not surprising that the ability to reach states of heightened consciousness improves as we age, because the ability to represent all types of information develops as the prefrontal cortex develops.

At this high reach of consciousness there is a heightened appreciation of the experience of being conscious – meta-consciousness. Meta-consciousness involves seeing yourself as a separate whole conscious person. In other words, we are, at our most conscious, aware of our consciousness. You are probably now having the somewhat vertiginous experience of meta-consciousness – it is like trying to watch yourself from without, or looking at yourself looking at yourself. Have you ever inadvertently stood between two mirrors and seen your image reflected to a diminishing tiny blob? If not, you can do so easily by putting a hand-held mirror behind you while looking into in your bathroom mirror. You can manipulate the hand-held mirror to reflect your own reflection in the bathroom mirror to smaller and smaller images. Your consciousness is like an infinity mirror system – your reflection goes from your real self to your image, and this image is then reflected in the opposing mirror, and this is then reflected in the opposing mirror, and so on. You see repeatedly diminishing frames containing the more diminished frames, to infinity.

When we say that we are looking at ourselves in a mirror we are not – we are looking at our reflection. Perhaps this is why the word 'reflection', as a process of thinking about memories and thoughts, is such an apposite one. We cannot physically see the whole of ourselves except through reflection in a mirror or through representation (a photograph, a portrait): if you look down you can only see from the shoulders down. The very thing that we are trying to conceptualize, consciousness, is what also defines our limitations. We are trapped within our own consciousness, within our own mirrors. It is only through a representation, a reflection, of the whole that we can see ourselves in full. We cannot *be*, that is, exist or experience, beyond these mirrors of our consciousness. This is why it is simply not possible for those with severe psychosis or those with profound depression or mania to appreciate that they are ill. One would need to be outside one's own brain to be able to see these disruptions of inner experience.

In the moment of consciousness the brain is in the process of *working* memory networks, and the integrative system of neurons, predominantly in the prefrontal brain, is processing the flaring inputs in the pruned lattices of learned organization to arrive at a pared-down output: the thought, the conclusion, the intuition, the prediction, the knowledge, the understanding. Without the integrative networks, input would not be assimilated, and the world would be incoherent. In working memory, all of the systems of representation in the brain that we have previously explored – sensation, motor and feeling – are integrated to form a living picture of oneself and the world in this present conscious moment.

The idea of connectedness, of being 'at one with the world', can be experienced in normal life. I experience this sort of transcendent consciousness when swimming in the sea. I and the other swimmers in our local bay go into the water, whether cold, or colder – the sea is never warm in Ireland. We get stung by jellyfish in the summer and beaten by cold easterly waves in winter, and tangled in seaweed by the shore, and return for more. There is only the sea and the sky as we swim out, nothing more. There is no gravity, only an immersion and a suspension, and we are live, integrated beings moving in the flow of something very big that is also alive. Connected with the elements – sea and sky – and completely self-directed, we pitch our bodies into the momentum of the sea and wind. At some point I begin to calculate how long more my body can take before the cold starts slowing me down, before I turn to swim back to shore. I emerge freezing and euphoric, feeling at one with myself, entire, and reconnected with the world. Sea swimmers share these moments of entirely private and common transcendence like an open secret.

Network Neuroscience

The late neuroscientist Gerald Edelman knitted together ideas of consciousness and memory that have resonated through to the most recent discoveries, and I will finish the chapter with some of his insights. Edelman was a molecular biologist who won the Nobel Prize in Physiology or Medicine in 1972 for uncovering the way that antigens recognize and remember immune cells.[114] During the latter part of his career, he looked at past and working memory, in particular how the brain 'recognizes' the present through memory. He saw this recognition as taking place in the conscious present, and that consciousness was the process involved in a re-working of the memory networks – what he called 're-entry'. Edelman intuited that there were multiple brain networks involved in any conscious experience and that these networks were forged by experience and re-worked by present input.

Today, Edelman's intuitions have evolved into what is called 'network neuroscience' and 'connectomics' – how the brain is connected up. Some of the work that young neuroscientists like Danielle Bassett from the University of Pennsylvania, also a physicist, undertake uses the principles of physics and mathematics to understand patterns in the neural networks that underlie learning and memory. In Bassett's work she 'matches' brain neural connections and activity, recorded through multiple techniques including fMRI and EEG, to particular subjective experiences and behaviour. Her graph theory has provided a novel method for understanding the patterns of brain disorganization in schizophrenia, where there is more random firing and fewer hubs of concentrated activity in the prefrontal areas. I see her elegant and precise search for brain maps of experience as being similar to Proust's introspective search to find past experience. Bassett and her contemporaries are thinking big, on the shoulders not just of the preceding giants of neuroscience, but also on the shoulders of our great introspective artists, who got there first.

12. Sex Hormones and Songbirds

While in the first two decades of life the brain is pruning and connecting up in patterns that reflect the learned exteroceptive world, great changes are afoot in the interoceptive world of feelings and emotions. After the sensate explosion of childhood comes the emotional implosion caused by the sex hormone deluge of puberty. In this chapter we will look at the brain changes that underpin romantic longings, sexual desire, partnerships and reproduction; and the memory systems involved in emotional learning and emotional regulation. What the developing adult will attend to and memorize will change for ever following the changes that sex hormones have on brain architecture. Nature provides a wonderful example in the male starling of what sex hormones can do in the brain, and how this changes behaviour.

Starling Song

I hope that you are as lucky as I am to be exposed to the beautiful sound of starling birdsong during summer. I once heard a spectacular volume of birdsong in late autumn but, glancing out the window, couldn't see immediately what must have been a massive gathering of birds. I then saw that the big sycamore tree in my front garden, almost denuded of leaves, was covered by a murmuration of probably more than three hundred starlings. They were perched on the branches, silhouetted by the backlight of a bright rose-orange sunset, and camouflaged to look like black leaves.

The cause of the orchestra of starling birdsong is that a 'song cortex' grows in the starling brain during the summer season. The song cortex develops because the sunlight triggers the release of testosterone in the male starling, and the testosterone latches on to testosterone receptors that are present in a part of the starling brain promoting the development of neurons that will eventually form the song cortex. The testosterone-induced song cortex grows with the increasing exposure to the light of the lengthening summer days, and their singing intensifies as the weather improves. This attracts the female starling, and courtship commences.[115]

When the days shorten, testosterone levels drop because of reduced exposure to daylight, and the song cortex atrophies in the male starling. With the demise of the song cortex, the starling song fades from the landscape. The female starling brain does not elaborate a song cortex because she does not produce testosterone, and so her songs remain very low-key compared to the courting male. If the female is given testosterone, however, she too will develop an elaborated song cortex. Cleverly, the testosterone receptors are present in both sexes but require the presence of testosterone to activate them and promote neural growth. The presence of both the hormone and their receptors is required. The brain is a-buzzing with life in which endless chemicals are floating in inter-synaptic gaps attaching only to their own specific receptors and activating them. This is an important point: while any sex hormones or neurotransmitters may flood the brain, they are only active where there are receptors for them to latch on to. I used to think that the starlings left us for the winter, but it is only their rich song that leaves us, to return again in the spring with the light-induced testosterone surge.

A similar sex-hormone-mediated process occurs in the human brain. The exact set of circumstances for triggering puberty are complex but are known to involve nutrition, genetics and also, in the case of girls, what they have experienced during childhood. In a fascinating body of research during the 1980s and 1990s, Jay Belsky noted that an absent father, but not an absent mother, was associated with earlier puberty in girls.[116] He posited that the explanation lay in human evolutionary history, when adverse conditions of survival, likely to be associated with an absent hunter-gatherer father, would necessitate earlier reproduction and independence in female offspring. There is a common physiological trigger for puberty, regardless of the multiple converging influences, such as weight and genetics, that impinge on the onset of puberty – a hypothalamic hormone, aptly named KiSS, or kisspeptin.[117] Brain hormones are made in the hypothalamus and released into the body where they direct further hormone release from different glands in the body. Remember the HPA axis, where the hypothalamic hormone CRH ultimately causes the release of cortisol from the adrenal gland? The KiSS hormone kickstarts the hypothalamus into producing a brain hormone that goes into the body to cause the production of male sex hormones called androgens, primarily testosterone, in the testes, and female sex hormones, primarily estrogen and progesterone, in the ovaries. The male and female sex hormones then latch on to their matched receptors to cause body and brain changes during puberty.

Importantly, in terms of brain function, the sex hormones are carried back in the blood circulation to the brain. Here again you can see how the brain directs activity in the body and vice versa. Within the endocrine systems, hormones produced in the hypothalamus bring about the synthesis and secretion of hormone in the glands of the body, and these hormones return to the brain to influence brain activity. This dynamic is analogous to the hypothalamus directing the ANS to cause body sensations, from which neural signals are returned to the brain to map emotions in the insula. What the sex hormones do in the brain is important in understanding what the developing brain attends to and memorizes. Alice was a patient of mine many years ago who only began menstruating in her late twenties. The effects of the monthly ebb and flow of the sex hormones on her adult brain created novel experiences for her, and her story remains for me a unique illumination of the mood-altering effects of female sex hormones on a sex-hormone naïve brain.

Alice

Alice had suffered from anorexia nervosa since her early teens, and her life revolved around the amount of calories that she ingested each day. She only ate food that she made from basic ingredients, weighing the ingredients in ounces, preparing all her food in a strict routine before she went to work each day. She ate with a discipline that was extraordinary. Alice kept her weight at a constant 38 kg (84 lb or 6 stones). Because of her consistently low weight she had menstruated only once before she developed anorexia at the age of thirteen years, and had not menstruated in the intervening years. A female has to be of a critical weight to ensure menstruation – this is probably a mechanism to prevent reproduction in the very young and to shut down reproduction in a situation of poor nutrition, where a pregnancy is not sustainable or an infant might die. Alice exercised with the same measured exactness as she ate: swimming X number of lengths of the pool Y times per week, and so on. There was almost nothing else in her life that engaged her. Although underachieving relative to her impressive abilities, she performed her work with the same meticulous care that she brought to her eating and exercise habits.

In an ideal healthy pattern of eating, food intake is regulated by appetite and terminated by feelings of satiety. We reasoned that this would not be a realistic goal, because Alice had not developed any knowledge of appetite and satiety, and she may always be beyond the regulation of more natural food intake mechanisms. We worked out a strategy in which, rather than trying to suspend her obsessional eating habits, we would use them to help her make modest incremental increases in her food intake. So, rather than going for the standard appetite-satiety-driven model, or the eat-to-a-target-weight therapy, we planned that she would use her obsessional patterns to regulate her food intake. She would, for example, add an extra one or two ounces of wholemeal flour in the making of her bread.

Alice meticulously adhered to her new regime, and very slowly, over a period of months, made modest weight gains. She suffered immense anxiety during her weigh-ins, often not sleeping the night before and requiring small doses of benzodiazepines to contain her feelings of panic. When she reached her target of 47 kg she began to feel strange and irritable. She felt as if she was out of control, and she was insistent that this was a feeling different from that of the anxiety related to her weight gain. She felt as if she had been colonized by an uncontrollable emotional virus. She became tearful, unusually sensitive and irritable, and she experienced, for the first time in her life, fleeting suicidal thoughts. Then she got a period – her first in adulthood, although she was in her late twenties. Her sex-hormone system had been kickstarted, as it is in adolescents, by reaching a critical weight, leading to the production of sex hormones and the activation of novel emotional experiences through hormone-induced brain pathways.

Alice was overwhelmed by her emotional instability and became unable to function. She was aghast that all girls and women went through these feelings on a monthly basis. We decided that a trial of antidepressants might help to control the emotional turmoil, and that she should try to ride it out with pharmacological and psychotherapeutic help. She and I were stunned by the effects that her hormonal awakenings had on her emotional systems. Her emotional brain did adapt, and we were able to discontinue her antidepressants about a year later. Following this orientation to the adult emotional and sexual world, Alice began to socialize. The following year she fell in love with a man, and I moved on to work elsewhere. She wrote to me the following Christmas with the news that she had married her man and was pregnant. She wrote to me a few Christmases later telling me that she had had another child. I moved on again and we lost touch.

Alice bears witness to the tyrannical world of anorexia. She triumphed over this with iron discipline and managed to manipulate herself to health by exploiting her own vulnerabilities. To return to the topic at hand, Alice was an adult observer of the mood instability that is an intrinsic part of normal puberty. Her story demonstrates the emotional experiences of all adolescent girls as the effects of sex hormones kick in. At first, the heightened emotionality brings a period of emotional turbulence, before improved socio-emotional skills emerge to allow courtship, partnership and reproduction. The sex hormones released during puberty contribute to many of the impulsive and emotionally uncontrolled behaviours that occur universally during adolescence and account for most of the mortality – traffic accidents, suicide and drug abuse. These emotional changes should eventually lead to new emotional awareness in oneself and in others. Once Alice found her feet in a new emotionally complex world, she went on to develop a more nuanced understanding of herself and others and to change the course of her own history.

The emotional changes caused by sex hormones bring about fundamental shifts in what a young person will attend to, how they will interpret it and how it will be memorized. An example of this, returning to the courting starlings, can be seen in the female starling response to the competing singing males. If she has low estrogen levels, she ignores the male songs and doesn't bother to select a mate. They simply do not have her attention – she is not aroused. When her estrogen levels are high, as they would be during the mating season, she will attend to their songs and choose a mate.[118] It was only when Alice's brain became estrogenized that she paid attention to courting men. The arousal of body and brain has life-changing consequences for the young adult. How many times have we heard romantic stories of the beginnings of relationships end with the line '. . . the rest is history'?

Sex Hormones and the Brain

The sex hormones latch on to specialized receptors that are present in specific brain areas, as with the song cortex of the starling. In humans, sex hormone receptors are present in prolific amounts in the memory and emotional hubs – the amygdala, the insula and the hippocampus.[119] Like the starling, we have estrogen and testosterone receptors in our brains regardless of birth sex or gender. Both the sex hormones and the sex hormone receptors in the brain are necessary to have changes in brain structure and changes in emotional responses and behaviour.[i] This is why the administration of male or female sex hormones to individuals of the opposite sex changes not only the external secondary sexual characteristics but also feeling states and behaviour. A simple example of this is that administration of testosterone to trans males generally increases sexual desire, while reduction of testosterone to trans females generally reduces sexual desire.[120] Both the trans males and the trans females have androgen and estrogen receptors from birth, but these receptors are only activated following androgen and estrogen therapy respectively. There is a rare disorder in which a person born with a male set of chromosomes – that is, XY, rather than XX – and who produces testosterone normally has non-functioning testosterone receptors. This is called 'androgen receptor insensitivity', and in spite of there being lots of circulating androgens they do not work. This leads to the person developing like a female sexually because the female is the default sex. Like the body, the brain also develops like a female brain, because the male and female brain, as with the starlings, develop differently under the influences of the male and female sex hormones. The brain of the XY person who has not been exposed to testosterone will be just like a female brain, with reduced amygdala responses to sexual images compared to the male brain.[121]

The hormonal effects on the brain can be seen in the higher emotional hubs in the brain, particularly the insula. In women, increased activity in the insula has been found to fluctuate in synch with levels of estrogens.[122] A human fMRI study from Germany that looked at how estrogens affect the emotional circuitry in healthy women demonstrated this nicely. The participant women, who had either high or low estrogen levels, viewed a movie with traumatic emotional content.[123] The authors reported that there was increased activation of the insula and the cingula in the high-estrogen women, compared to low-estrogen women. In this way the sex hormones, through activation of hippocampal-amygdala and insula circuitry, modify what we attend to, how we feel and what we memorize.[ii]

The Prefrontal Cortex and Emotional Development

If the years of emotional disquiet in adolescence can be managed, most young adults will emerge into states of improved emotional control. This occurs because of development in the prefrontal cortex that facilitates emotional regulation, in keeping with the fact that the prefrontal cortex is an integrative region where thinking and feeling states are processed consciously.

The famous and unfortunate case of Phineas Gage from the mid-nineteenth century is taught to all medical students when learning about socio-emotional aspects of prefrontal function. He was a working as a railway construction worker in Vermont when a horrible accident resulted in an iron bar being blown into his head in an upward direction from under his cheek and coming out through the top of his skull. He did not die immediately but, astonishingly, was conscious and orientated within a few minutes. He went on to regain motor function, but his personality thereafter was fundamentally changed. Before his accident he had been an unremarkably agreeable man; following the accident he became callous, rude, socially inappropriate and unpredictable. This was one of the first clear demonstrations that the brain's social and emotional control centre was located in the frontal lobes of the brain, the region selectively severed in Phineas Gage. Severe frontal lobe damage usually changes personality, and the individual becomes more socially crude, disinhibited, indifferent to others and apathetic.*

I once saw a young man who had tried to kill himself by shooting an arrow through his head. The arrow went through his eye and out through the top of his skull. We were bewildered that he denied feeling depressed and that he was so blasé about his situation. I remember him telling us that he didn't care about anything and that he tried to kill himself, not because he was emotionally distressed, but because he had no interest in life. Young adults can present with 'boredom', but this is usually an expression of their inability to enjoy life. This young man had something different – a sort of pathological ennui. I had never seen someone who had used such lethal methods to kill themselves and were not depressed, nor have I since. Because of this, his case has remained with me. I often wonder if the arrow might

---

* A less well-known fact about Gage is that his social skills improved over time.

have changed his frontal lobe function, as it did with Phineas Cage, and made him apathetic and indifferent, as happens frequently when the frontal lobe is damaged. Had he been depressed before the suicide attempt, and did his presentation change because of the brain damage? Did he perhaps sever some connections to his prefrontal cortex, in other words, give himself a lobotomy? [iii] Once better from his physical brain injury, he was discharged to his local psychiatric hospital for observation and treatment, and to my regret I did not follow up on what happened.

Joseph LeDoux is a neuroscientist from New York who, when off duty, fronts a band whose name, The Amygdaloids, will give you an indication of his research passions. He offers some interesting insights into the processes of prefrontal emotional development. His research has demonstrated how connections from the prefrontal cortex to the amygdala and to the insula are critical to developing emotional control.[124] Part of the prefrontal development that occurs during early adulthood involves strengthening connections between the prefrontal and the emotional networks. LeDoux makes the profound point that, 'Contrary to common wisdom, extinction [of uncontrolled emotional states] is not similar to forgetting but instead represents new learning.' The prefrontal circuits that grow down to the amygdala lead to a growing *inhibition* of the amygdala output, giving a less intense amygdala hammer-gong of emotional experience and more measured feeling states. Inhibition is new learning. This can be visibly seen in increased brain activity in the prefrontal–hippocampal–amygdala systems.[125] Developments in memory systems of self- and other-awareness allow the developing adult to see things in perspective and not to react impulsively. In psychiatry and neurology, and increasingly in common language, it is said that emotionally disinhibited people are 'frontal'. The way that this term is used is counterintuitive, for me at any rate, because the person said to be 'frontal' is really frontally impaired, or *hypo*frontal.

Gender identity or sexual orientation may change, one's affiliative allegiances may switch . . . everything is up for grabs during the biological upheavals of emerging adulthood, depending on your genes, your environmental stability, your match to your environment, your childhood and the interactive effects among all of these factors. We are inimitably complex in our eccentric mixes of genes and development, and we are wiser to note the various influences and the amalgam of these influences for each individual rather than trying to generalize and weigh one against the other. Genes sometimes trump environment and vice versa: it is all a mix. We can be the emotionally balanced inheritors of experiences during development that have been directed in a caring and contained adult environment, or the emotionally uncontained victim of a chaotic environment during development. One may have kind and emotionally containing parents but be the unfortunate inheritor of genetic processes that unfold during brain development in adolescence, leading to psychotic or mood disorders.[122] Yet other adolescents have to negotiate the interoceptive disquiet of puberty in bizarre systems of social organization. Saddest of all are the individuals who have to unlearn responses to an abusive world and reformulate foundational memories to enable healthy maturation of personality.

The disruption of adult emotional equilibrium because of childhood abuse is most evident in the clinical diagnosis of borderline personality disorder (BPD). Abuse and neglect, like all biographical events, will modify brain architecture and memory pathways. Early life adversity results in an increase in most psychiatric disorders in adulthood – depression, anxiety, substance abuse, psychosis, death by suicide – but BPD is most specifically related to childhood abuse and neglect.

Borderline Personality Disorder (BPD)

BPD is a stigmatized disorder characterized by high levels of emotional distress and anger, rapidly fluctuating moods, a poor and shifting sense of identity, intense but unstable relationships, frequent substance abuse and recurrent self-harm. This pattern of feeling and behaving frequently follows upon childhood abuse or neglect, in which the child cannot learn to contain their emotions because they have not had the stabilizing influence of an emotionally mature adult. Parents need to initially contain the infant's emotions through soothing: the parent provides the soothing that then allows the infant to develop skills to self-soothe. Following infancy, when the child is developing the skills to learn through reason, they need to be guided behaviourally and verbally to contain their emotions, or they may become emotionally dysregulated adults.

Martin Teicher, who works in Harvard Medical School, has written that the brain alterations that occur in those who have been abused 'are adaptations to an anticipated stress-filled malevolent world'.[126] Anticipations and predictions are based in childhood experience, and while hostility and anger may be appropriate responses for a child who is being abused, they become self-defeating and maladaptive in adulthood. A person becomes overly defensive because they have been abused, isolating themselves further from the human contact that they have been denied and crave. There is evidence that the changes in brain systems brought about by abuse are probably specific for the type of abuse and for the timing of the abuse in relation to brain development. Verbal abuse or witnessing violence at home seems to cause changes in the auditory and visual cortical pathways, that is, the sensory memory pathways.[127] Emotional or physical neglect results in differences in amygdala-hippocampus, and in the higher integrative brain involved in emotional regulation – the cingula and the prefrontal networks.[128, 129]

Neuroimaging reports of individuals who have a diagnosis of BPD show that, in keeping with their experiences of intense and uncontrollable emotional states, their amygdala reactivity to cues is prolonged.[130] The prefrontal inhibitions that should grow down from the prefrontal to the amygdala-hippocampal region seem not to happen in an average way. As you might expect, given the absence of the brake of the prefrontal inhibition, the person with BPD has poor impulse control and will frequently abuse substances.[130] BPD is like living in an arrested state of emotional development. All of the neurodevelopment that should proceed – impulse control, social manipulation, emotional regulation, identity formation – gets halted somehow. It is natural for emerging adults to feel emotionally unstable and to have poor impulse control and faulty judgement, but in BPD these feeling states and behaviours are ingrained and outlast young adulthood.

It can seem sometimes to those working in psychiatry that individuals with BPD never learn, but research and longevity of experience in clinical practice indicates that this is not the case. Inner feeling states and measured behavioural responses tend to improve with age.[131] It has been found that those with BPD respond most effectively to a psychological therapy called dialectical behaviour therapy (DBT). DBT is based on the dialectical position that one has both to accept oneself as one is, while simultaneously moving towards change.[132] This therapy was founded by Marsha Linehan, who herself suffered from BPD for which she was hospitalized for many years when young. It is similar, I often think, to parenting for adults. I am not suggesting that all people with BPD or traits of BPD have been poorly parented, because some people are simply more emotionally unstable, have poorer impulse control, are more likely to become addicted to particular or any substances, experience more anger, or possess a weak inherent sense of themselves as an exigent human in the world. Research has shown that there are innate personality traits that seem to be constant over time, and stability in these traits leads to easier maturation.[133] It is not always related to parenting.

In DBT, behavioural boundaries are negotiated, such as not self-cutting or threatening suicide, and the person is simultaneously given intense support and insight-oriented psychotherapy. Similar to firm parenting, this allows the person to feel safe within the boundaries of agreed behaviours while they learn about emotional regulation. LeDoux's model of prefrontal inhibition as representing new learning, and the success of DBT in modifying emotional states, all lead to hope for healthy change in those with BPD and for those whose memory and emotional networks are skewed by adversity. A neuroimaging study in individuals who underwent DBT has demonstrated that there is growth in inhibitory pathways from the prefrontal cortex to the amygdala with improved emotional regulation.[134]

All things going well, and in the absence of early life adversity, the prefrontal neurons grow down during early adult life to inhibit amygdalar output, and an emotional equilibrium is reached. This shift allows a stable sense of self to emerge that forms the foundation for a stable personality and identity. Understanding the emotional experiences and intentions of oneself – emotional insight – and of others, improves with age and in line with prefrontal developments.[121] Emotional maturity is more likely to lead to a successful and happy relationship than the romantic yearnings of youth when sexual attraction is not balanced by experience. This is not the only advantage to being older in the romantic domain. Although we are constantly being given the message that a youthful appearance will make one more attractive, this is not always the case. We'll finish the chapter by returning to starlings, who provide a wonderful example of how memory and learning can confer advantage in the romantic arena.

Starlings, similar to humans, are socially cooperative but are also competitive at an individual level. They travel in huge broods, murmurations, undulating in the sky in swarms into beautiful changing rounded shapes. And there is no leader, just their brilliant radar that keeps them perfectly spaced from the surrounding starlings in flight. When it comes to breeding, they are less communal, and once their testosterone-fuelled songbird cortex is primed by sunlight the males stand at the entrance to the buried nests competing with other males to attract the resident female. Female starlings will go for the best singer, and the best singer in the female starling's world is the one with the longest and most complex song.

Because starlings are brilliant mimics, they learn songs from listening, building up a varied repertoire as they age. This gives the advantage to the older males, who, because of their song memory, win the females. Meanwhile, the young male starlings learn new tunes from their elders, creating melodic memories that will give them the advantage for breeding seasons to come. Reassuringly, it is not only youth and beauty, but experience and memory, that brings about romantic success.

13.   The Shifting Narratives of Life

This is what fools people: a man is always a teller of stories, he lives surrounded by his stories and the stories of others, he sees everything that happens to him through them; and he tries to live his life as if he were recounting it.

Jean-Paul Sartre*

In preceding chapters we have looked at the neurodevelopment of integrative memory systems in the brain and how this leads to a point where understanding of the exteroceptive world, and of oneself and others, is possible. Essentially, prefrontal development allows representation of complex sensory information that can then be assembled to create a coherent story of events that, if biographical, probably involve the hippocampus. Pruning and myelination of memory networks in the prefrontal areas allow this to happen, and as the networks evolve we learn to predict, to imagine and to create. We have examined emotional neurodevelopment and how a young person learns to become aware through mirroring of themselves and others and to regulate their emotions through prefrontal growth, leading to some sort of equilibrium with the world. The journey to the formation of a stable way of being in the world is a life-long one that is always shifting as the present changes, as it has to, and new events and insights modify existing memory networks.

During some periods of history there have been immense and lasting changes in the world order – during wars and plagues – that required individual changes in memory networks. Boris

* Jean-Paul Sartre, *Nausea*, trans. Robert Baldick (Penguin, 2000), p. 63.

Pasternak's novel *Doctor Zhivago* is set during the years of the seismic societal changes of the First World War and the Russian Revolution. The novel is as much about individual as societal change: 'Everyone was revived, reborn, changed, transformed. You might say that everyone had been through two revolutions – his own personal revolution as well as the more general one.'* If the function of biographical memory is to make a coherent story of one's life with the time, place and person coordinates of the neural cell-assembly lattices, then the story being spun must be a shifting one. In this chapter we will look at how the individual, as Henri Bergson wrote, creates 'oneself endlessly', or how we narrativize our lives. Although the words 'story' and 'narrative' are used synonymously, narrative goes beyond the bare bones of a story. A story will always form because human neural networks will assemble into patterns: that is what they are designed to do. But we usually go beyond this to ascribing meaning to the story. Self-narrativization best describes the process of ascribing meaning to one's own experience and one's life story. Self-narrativization often involves the ladders of vanity that the older Yeats had to climb down from in older age, the ladders that held him aloft from the rag-and-bone shop of his heart, where we will all eventually lie down. We do this too in social and cultural memory, which we will look at in the next chapter.

My first experience of changing self-narrativization was a vicarious one, when I read Jean-Paul Sartre's novel *Nausea* in my late teens. You could say that the novel is simply about a young adult individualizing in a world that does not make sense to him, and so he feels alienated and disconnected. At the time I did not see this, and I was blown away by the power of the protagonist's visceral experiences and how they seemed to represent his intellectual struggles. Having recently re-read it I am even more astonished at its scope and originality. Although not really being able to understand it years ago on my first reading, I identified with the

---

* Boris Pasternak, *Doctor Zhivago* (Pantheon, 1997).

protagonist, Antoine Roquentin. The interoceptive depth of the novel hit me before I had any intellectual insights into how feeling states are a part of how we integrate and memorize the world. Roquentin is 'every' young adult person, disaffected and dysphoric, and he is also the embodiment of the upheavals in society at the turn of the twentieth century, as awareness of new levels of consciousness were bursting though.

The book gets its title from the unpleasant interoceptive feelings, in particular the nausea, that the protagonist experiences. Roquentin was living in a world to which he felt no affiliation, by which he was viscerally repulsed. He not only sees others as alien, he experiences *himself* as alien. Through his unpleasant hyper-perceptive experiences, the reader feels his morbid depersonalization. It feels as if the stream of present sensation from the external world is disconnected from his memory, leaving him adrift in a shifting and capricious consciousness that has no coherence. The reader is told that Roquentin had lived a normal life of old certainties, one of contented connection to others: 'I was inside it [my life]; I didn't think about it.' His world slowly falls apart leaving him 'rejected, abandoned in the present . . . '.* I see it as if there was no neural fit in his memory to process incoming sensation.

One of Sartre's achievements in *Nausea* for me is that he disassembled the building blocks of consciousness – the coherence of a continuous sense of self that is the outcome of integrating present events with memory. Biographical memory knits the present to the past and future giving us a sense of 'being inside' our own lives. Roquentin is living in a continuous, suspended present, imploding with sensate experience that he is unable to process. He resolves the crisis by deciding to create a new life, moving from provincial France to a new life in Paris. He would henceforth 'build [his] memories with the present'. Sartre and his fellow existential travellers, notably Simone de Beauvoir as a first-wave feminist, individually left their own lives of lived-in certainties to bravely create a new world view.

* Sartre, *Nausea*, p. 61.

This book resonated with me and some of my young adult peers in the Ireland of the 1970s and 1980s probably because many of us were leaving a set of discredited belief systems and the stability of a shared cultural identity. This did not bring with it the liberating sense of freedom that I, at any rate, was expecting. It often felt strange and disorientating, as described by the Roquentin of 1938. Everything was blown apart before it could be reconstructed. The experience of constructing one's sense of self, one's identity, may be more difficult when society is in rapid change, but it is always challenging for young adults. Life-as-lived in any culture generally guides one, initially unsteadily, into conceptual frameworks for understanding the culture and one's place in it. For some, like Sartre and de Beauvoir, there was a dissonance between how they experienced themselves and how they understood the world that they ultimately could resolve only by changing the world.

Trauma

Sartre's book is so deeply embedded in my memory that I was jolted when my patient Araf – the clever young man with mania – told me, paraphrasing the character of Roquentin, 'I have to rebuild what I learned about myself,' followed by, 'It is only human to change memory.' In most of the individuals that I treat there is a dissonance between themselves and the world, but not usually because of existential angst or seismic social and political changes, but because of childhood trauma and/or serious mental illness. Trauma often requires the individual to create an almost new self and a new set of memories. Frances is a patient of mine who was ensnared by an enmeshment of toxic childhood memories and psychotic misinterpretations, and she has given me many insights into how trauma and/or psychosis can lead to a monstrous self-narrative that can destroy someone. It is important to establish that individual trauma, such as death and illness, does not have meaning in itself, and that it is dangerous to ascribe meaning to trauma, but we'll return to this. Frances, as you will see, is a remarkable person, and hers is a rich story that goes back to the days of the old asylums.

Frances

Frances's childhood had been very bleak indeed. She was born the third child in a sibship of four. Their father was alcoholic and abusive, frequently physically violent, and although he had a steady job he drank his wages. Mother spent most of her time outside the home looking for food and means to support the children. All the children were neglected physically and emotionally. Frances was sexually abused by her father and by men, traders and visitors, who came to their home. She often did not attend primary school and was generally left to her own devices during the day with her younger brother, the only person with whom she had a loving relationship. She had no peer friendships and no confidant. Her attendance at school became less frequent as her childhood progressed until she stopped going altogether. There is no record of her attending secondary school.

She frequently ran away from home. When she was eleven years old she started living on the streets, learning to survive by shoplifting and begging. She slept in doorways and phone boxes. Occasionally the police would pick her up and her mother would come to the police station to bring her home temporarily. Street life exposed her to further violence and abuse. She lived according to whatever means she found and learned how to fight and to defend herself. When she was thirteen Frances found shelter with a fortune-teller for about one year. She had fallen into a pattern of addiction to substances, mostly alcohol, by her mid-teenage years. She was sixteen when she was first admitted to a psychiatric hospital in Dublin city.

A brief diversion from France's story is necessary to give you a sense of this hospital. The hospital was colloquially known as Grangegorman, and if that name brings to mind Mervyn Peake's Gormenghast you are getting the picture. It was a prototypical Victorian asylum: a big granite structure, grey and desolate and crumbling. I asked to be placed in Grangegorman for a six-month stint during my training because it was one of the few remaining old asylums in Dublin. It was a very grim place, almost derelict in parts. At this point in its history patients were being moved out to live in the community and Grangegorman was in an accelerating state of disrepair. There were endless long corridors, with rows of rooms on one side that looked like prison cells facing large barred sash windows on the other side of the wide corridors. Some of the residents would rest on the deep, long window sills with their back against one set of folded shutters, which had not been opened in years. Their feet rested on the opposite shutter of the ample sills. Once when I was walking one of the long corridors at night when on call – lit only by outside light coming from the huge windows – a patient darted out from one of the window sills frightened at having been woken from sleep.

I learned over the six months that the asylum was more complex and less harsh than the stereotype that I had expected. I was full of the predictable indulgent indignation when I started . . . the residents had had their agency taken away by a cruel system that was centred around institutional efficiency rather than the needs of individual patients . . . psychiatric institutions were inhumane . . . patients were objectified and wandered around in circles, sedated and drooling. Some of this was true, but the system of care was more layered than the abject stereotype of the Victorian asylum.

One event from this time sticks in my memory. An older consultant brought me on a ward round one day that took us to a hidden quarter of the hospital. This island of habitation was situated within a bigger, almost abandoned, granite building. We entered through a door in the gable wall and went up a few flights of stairs. This led to a big, open, Florence Nightingale-type ward, with two long rows of beds, separated by curtains, lined up on opposing walls. Most of the room was taken up by a big central open space. Someone jumped up from a bed on our arrival and hurried over to us: to my surprise it was a male nurse. The other occupants of the beds were lying fully dressed on their tidy beds waiting for the consultant's regular weekly visit. They rose from their beds when we arrived in the ward. He spoke to his patients like one would to a fellow villager, passing the time of day with pleasantries. There were no queries about hallucinations, delusions or medication. I followed him around asking him about the different illnesses of the individual patients and their medications. He more or less ignored me in an affable way.

When these pedestrian exchanges were concluded with the patients, I followed him, a bit bewildered, down a long corridor that emerged from the other end of the ward. This corridor spanned the length of the building, having the typical cell-like rooms on one side and the large windows on the other. In one of these rooms lived a man who was a watchmaker. It was the time of mechanical, wind-up watches, and every surface of his room was covered with the paraphernalia of his trade: cog wheels, cases, glass covers, chain straps, oil cans. The watchmaker fixed everyone's watch in the institution – patients and staff. They had a chat about the consultant's watch that he had fixed, as I'm sure they did every week. In the early years of his illness the watchmaker had been a 'revolving door' patient in Grangegorman, being admitted again and again believing that his family and neighbours were trying to kill him. Finally, it was decided that it would be best for everyone if he stayed in the hospital.

Looking back on those six months I remember the corridors and the watchmaker's room and now can see that the old asylum system was a flawed form of care, but not a cruel one. We have swapped it for a system that almost excludes residential care, and patients and mental health workers, instead of struggling with the restrictions of institutionalization, struggle with poverty, homelessness and the criminalization of behaviour resulting from psychosis. Over the years I have come to appreciate the style of the older consultant from Grangegorman, and the power of simple, normal human inclusion. The asylum system, when functional, provided a protected village. So-called 'community care' has brought vulnerable psychotic people into the outside world, where there is often no asylum. The whole community psychiatry concept was based on the assumption that everyone with a psychiatric illness could be brought to a level of functioning compatible with the open world, and that the world would have sheltered niches for them. Psychiatry was wrong on both counts. As far back as 1939 the psychiatrist and mathematician Lionel Sharples Penrose noted that there was an inverse numerical relationship between the number of psychiatric beds and the number of prisoners. This remains true today.[135] Whatever city one travels to, however affluent or however functional the healthcare system, however respectful the state claims to be of individual rights, the psychotic are on the streets living rough, being ignored and neglected.[136] They are too ill to negotiate social welfare systems, and when they become too much of a public nuisance they are jailed. This demonstrates again that, despite the improvements in public awareness and concern about mental health, this has not permeated to the mentally ill who have brain diseases.

Grangegorman is where Frances spent her later childhood years. Following her first admission until she reached legal adult status, she had about twenty short admissions. There were no inpatient treatment facilities for children or adolescents in the 1980s. When not in the psychiatric hospital, she seems to have been in prison, on one occasion serving a sentence of four months for drunk and disorderly behaviour. She was seventeen years old at the time. Thankfully, some of her clinical notes from her time in Grangegorman followed her to our service. They are illuminating, not least because of the kindness that her psychiatrist, Dr F, revealed in his clinical notes and summary letters. His care and understanding of her situation and his efforts, and those of the nursing team, to provide respite from the horror of her developmental years shines through.

When she was eighteen Frances disappeared from the psychiatry services, and for the next eight years we have no notes on her. This coincides with her meeting Kieran, with whom she fell in love, and he 'took me home', as she put it. She wound up back in a psychiatric inpatient unit in her late twenties. In the intervening years, Grangegorman had been closed down, staff had been transferred to suburban hospitals and patients had been transferred to what was being uncritically lauded as a more humane system of care in the community. As life and luck would have it, Frances met with Dr F again, who became her treating psychiatrist. This was fortunate, because Dr F was now able to elicit descriptions of psychotic experiences that Frances would or could not volunteer during her teens. Over the next few years her course fluctuated, but there was a pattern: she seemed to go from the florid psychosis of hallucinations and delusions to a less psychotic but self-destructive state. It seemed that when the terror of the psychotic experiences receded, another terror rushed in. This led to a pattern that is referred to as 'sabotaging treatment'. The explanation, I believe, is not that a person doesn't want to be well, but that they don't understand how to live without their psychosis – they have no normal memory networks. Frances had never processed real-world experiences, and she would have to learn to do this much like a newly sighted adult if exposed to a flood of visual images. Over a long admission she appeared to connect with the outside world in a tentative way and was finally discharged.

When I took over her care about two decades later, her life, on the surface, appeared to have settled. Over the years she had done a rehabilitation course in art, and this led to her successfully completing a degree in fine art. She was attending psychotherapy with an experienced psychologist and was abstinent from drugs and alcohol. She had only self-harmed during times of significant stress. On our first consultation, she was dressed carelessly in multiple layers of clothes, her head was lowered and she only made darting eye contact. She told her story to me in a very low voice, automatically, as if she had told it hundreds of times, seeming to be disconnected from the emotional experience of traumatic past events. She was also immersed in psychotic beliefs and experiences, believing that she felt other peoples' emotions in the concrete sense of individuals transferring their experiences into her body; 'I can take on other people's emotions . . . on their behalf.' The confusion of self–other boundaries happened routinely, particularly if she became close to someone. The converse, that someone else was inhabiting her, was also happening, and, in her case, it was 'the devil'.

I got the impression of someone who was intensely introspective and that her inner world consisted of a confusion of psychotic ideas. I was *unsettled by her biography, having rarely heard a personal history of such grim pervasive abuse. There had been no redeeming adult figure in her childhood, no reprieve from the unrelenting abuse, and no social stability apart from the Victorian asylum of Grangegorman. Somehow in all this chaos, but exactly when was not clear, she had developed a serious psychiatric illness. I wanted to spend at least some time contemplating what she had told me, to pay some tribute, if only for a few private moments, to her unimaginable young life. But a waiting room of patients in a public psychiatry clinic does not allow for even such a modest tribute.*

Her main problems at this time related to her reclusive behaviour. She spent most of her days lying in bed, immersed in psychotic experiences in which her abusive past was intruding into her present consciousness. Everything that she could possibly experience outside of the care and love provided by Kieran, and the sanctuary provided by the psychiatry services, provoked toxic memories. It was less painful to live in a world of sensory and emotional deprivation, reclusive in her room, than to constantly experience the continuous toxic memories provoked by input from the world. I altered her meds to try and better manage the psychosis, and we implemented a programme to try to bring Frances into a sheltered social world that would be less threatening, hoping that we could gradually shift her paranoid interpretations. We coaxed her into attending our day hospital so that she would have a daily structure in a social environment – a sheltered microcosm of normalized human encounter, along with individualized therapy. The journey into her past, and how this had formed her present perceptual filter, was negotiated with skill and sensitivity with our psychologist over several years.

Sometimes, psychosis has been the individual's world for so long that they don't want to leave the familiar. Do any of us? Individuals with psychotic experiences, even severely grotesque experiences, may be terrified of leaving this behind, because they could be exposing themselves to the threats that still remain present but are now invisible to them. Sometimes they are comforted by their voices and feel lost without them. Sometimes they do not want to leave grandiose beliefs that they have of their special powers. Sometimes acknowledging chronic mental illness is too painful. Only very rarely has a patient confided to me when free from psychotic experiences that they want their psychosis back, but I suspect that this happens much more than we realize. It is the individual's choice to refuse treatment and to be psychotic, unless they become a threat to themselves or to others because of their persecutory delusional beliefs. In this latter situation, psychiatrists are legally obliged to treat, involuntarily if necessary and regardless of the person's personal wishes.

Individuals who have been psychotic for long periods of time require a decent interval free from psychotic experiences to test out the 'shared' world of common experience and adapt to it. Following the resolution of the immediate experiences of psychosis, the provision of a safe world in which to create new networks based in a shared common reality, to 'build memories in the present', is just as important as the pharmacological treatment of psychosis. Our team in the day hospital became Frances's social world, her village. She contributed to the life of the hospital through her soft personality and her art. Trusting relationships with the treating team members provided a helpful challenge to Frances's belief that she was not able to manage relationships outside of that with Kieran. Over time she developed warm relationships with us and with other patients, which sustained her, even through the later tragedy of Kieran's death. We were also enriched by our relationships with Frances, who engaged us with her individual combination of stoicism and vulnerability, and her emerging wry sense of humour.

Trauma

Frances's memory networks had been formed, to repeat Martin Teicher, as *adaptations to an anticipated stress-filled malevolent world*. Her brain had been wired to survive in a hostile, and not a benevolent, world. One of the most basic necessities of survival, either individually or as a species, is the ability to adapt to one's environment, even if this is in a world of pervasive abuse. Humans have immense social adaptability – we leave our families in young adulthood and move to peer-group identification, before moving usually to monogamous relationships from which we may detach in the future, or not. We form new bonds and attachments, loved ones die, we grieve and adapt. Frances had adapted to a profoundly disturbed world and was deeply traumatized.

Trauma is defined in DSM 5 as 'actual or threatened death, serious injury or sexual violence', all of which Frances experienced during her childhood. In less extreme circumstances, life trauma may result from happenings that are not life-threatening but lead to such emotional distress that they cannot be integrated emotionally. In the language of psychotherapy, the experience cannot be emotionally digested. Following trauma, time seems to freeze into a few scenes that are played and replayed with the hammer-jack emotions of amygdalar circuitry. The past is intruding, reverberating, in the consciousness of the present. Nick Cave, the Australian songwriter, who sings melancholy as if he is playing tunes on the listener's insula, has put this beautifully in a documentary about his son's death at fifteen years of age. He says that his son's death is like an elastic band. He can move and stretch into the present, but at a certain distance of stretch he is pulled back. What is called 'moving on' can sometimes feel impossible.

While a traumatic event is unique to an individual, the replay with intense associated emotions and the inability to 'understand' is universal. I have heard the phrase 'I don't understand' so often from traumatized individuals. I've heard it from a soldier who returned from war having witnessed one child soldier kill another, from a woman who delivered a stillborn baby having felt it kicking *in utero* pre-delivery, from a parent whose teenage child died from suicide, from a mother whose son lost his life to a casual act of violence. It is as if there are no prefrontal groundwork lattices into which the event or events can be integrated, and this has to be built, one painful connection at a time, while others are withering away. The intrusion of grief into present experience reduces as the memory networks shift and slowly the person moves forward to inhabit the present. Do we sometimes, somehow, in our grief and love following loss, try to inhibit the changes in network configuration that occur inevitably with being alive? Is it unconscionable to put the memory to rest? Does the bereaved person feel that without the dead they would become an even more alien intruder in the now estranged and seemingly carefree world that they are so still within? Is it resisted as a sort of conscious trade-off for the feeling of the departed in the present, however painful? Whatever and whichever, trauma 'takes time'. Edith's flashback memory that we opened with shows how the original memory of the traumatic psychotic experience had been activated and vividly replayed. But this experience, the recall, was probably re-stored, on the second occasion, less proximate to the emotional experience of the psychosis. With each future sighting of the gravestone, or imagined sighting, the psychosis would be one more step removed from the image, until all that might remain on seeing the little gravestone again would be a sense of feeling disturbed. This is not a repressed memory, it is the best that can be done with traumatic memories – it is 'resolution'.

Frances also taught me the importance of what may seem to be casual social exchanges. This is sometimes the only communication that individuals have with the social world. In the world of damaged memories and over-sensitized interpretation, the quality of simple human interchanges cannot be overestimated. Most conversations in Ireland start with comments about the weather; there is an inestimable amount of emotional and cultural communication exchanged in what could appear to be formulaic exchanges. It is, more than anything else, our shared reality – there is great joy if we have a warm sunny day and much sharing of misery on rainy and windy days. The stereotypical, hopefully fading, idea of the psychiatrist cleverly breaking down speech content over hours of uninterrupted non-directional therapy could not be further from the workaday reality of the average clinical psychiatrist. We are more healers than people who break down, more interested in coaxing our patients back to the shared world than trying to plunge them into interospective chaos, more practical than theoretical.

In this process of coaxing a chronically psychotic individual back to the shared world, exchanges are kept simple and unambiguous. The mental health gains in a person who has lived with long-term untreated psychosis and their participation in the world may sometimes seem modest to an outsider, but it was a triumph for Frances to have a chat with a nurse about the weather, the journey to the day hospital, her new dress or the price of cigarettes.

I now think that we are all like Frances to some extent, living our lives in an equilibrium between a potentially threatening outside world and our sometimes sensitized memories. The crucial issue is to continue to try to develop a more healthy equilibrium. Frances is not raging against the world, she is not destroying herself or the world around her; she is endeavouring to participate. Frances and other patients with psychotic disorders have taught me that the key issue for a human being is not to be highly functional on an abstract scale of functionality, but to be in equilibrium. If there is anything that I have learned from my work with mentally ill patients it is that the achievement of an easy equilibrium between oneself and the world is what determines one's happiness. I was struck by a phrase in a Michael Cunningham novel, *A Home at the End of the World,* in which he describes a group of psychologically fragile individuals as being 'held together by duct tape'. They evolve to live as an eccentric ensemble in a houseshare, negotiating an emotionally delicate dance around each other with mutual laconic sensitivity. Not necessarily happy, and certainly dysfunctional, they find their own home, albeit one at the end of the world. We all need a home in the world, wherever that might be. One's state of normality, as Hippocrates would have called it, is for each of us to find, and none of us to judge.

14. *False or True?*

'I have done that,' says my memory. 'I cannot have done that,' says my pride, and remains adamant. At last – memory yields.

Friedrich Nietzsche*

In 1899 Madison Bentley conducted studies in which he observed and recorded the recall of simple images of coloured cards. He found that the accuracy of the recall of the first colour presented was reduced by the presentation of subsequent colours. The recall was made fuzzy by the competing sensory input that followed. In this, Bentley showed how subsequent memories change previous ones, an observation that may seem banal and obvious, but is actually neither. We may intuit that our past drives the present, but the events of the present also change past memories, as we've explored. Present experience and memory are in a never-ending dance of construction and reconstruction. Bentley's studies reflect how concepts of memory had shifted from one of fixed impressions, like the wax imprints of the ancient philosophers or the mechanical concepts of Descartes in the seventeenth century, to an organic associative process involving sensory experience, physiological arousal and emotion, by the end of the nineteenth.

Bentley's experiments were superficially modest, but ideologically sophisticated. He is a largely unsung pioneer in his meticulous observations of memory and his clever intuitions. We now know that we do not have a fixed store of memory that may or may not be available

* Friedrich Nietzsche, *Beyond Good and Evil*, trans. Marion Faber (OUP, 1998), p. 58.

for recall and that new input from the world is not simply augmenting established networks, but that there is a plastic network of connectivity between present input and memory. Even a new colour, if quickly presented, will shift the cell assembly that represented the previous colour, and this will be disrupted again if a further colour is quickly presented. This principle holds in relation to the more organized and complex process such as biographical memory. The hippocampus is engaged in putting together incoming sensory experience to create new biographical memory, but this putting together of current cell assemblies with existing prefrontal lattices will of course change the pre-existing lattices. One may be aware of the processes of working memory – thinking, actively recalling or imagining – but we are always silently processing information, what I think Baudelaire meant when he wrote about *la parnasse fecund* – 'fertile laziness'. Neural currents are always at play, swirling about in response to the zaps from the body and the world, regardless of whether the brain process is one of intellectual concentration or of fertile laziness.

The Bentley paper is a delight to read because his simple observations have remained consistent with the unending complexities that neuroscience continues to reveal, and because of his lyrical writing. He also found, incidentally, that bright colours are more easily remembered with accuracy. This is because bright colours cause more physiological arousal, and when we're aroused so too are the brain neurons, being pitched to fire and flare into their neighbours and forge cell assembly memories. Perhaps this is why the sun seems to have always been shining during childhood. In the final paragraph of the paper he wrote, 'From this standpoint are explicable the partial passage of memory into phantasy, and the weakening of memory fidelity.'[i]

Biographical Memory

If simple sensory cortical memory, like colour recognition, can be so easily manipulated, what are the possibilities for biographical memory? The only thing that is certain about biographical memory is that it has to change – *an event only occurs once*, as James Clerk Maxwell told us. Although there is little fidelity in biographical memory, as we've seen, in this infidelity there are many degrees of betrayal. On one pole there is fibbing, when one is deliberately telling a lie. Telling lies was a sin in my day, and when we trooped into the confession box to confess our sins to the priest every month we all said the same thing: 'I have told lies and have been disobedient to my parents.' Even then it was seen as something benign. This may sound uncomplicated, but a deliberate lie, or an obfuscation, can transmute into something more like an ambiguous memory, which can then become more like a real memory.

Author's Miller's play *The Crucible* is a clever exposition of how deliberate lying can transmute into belief. The play centres on the Salem witch trials that took place in Massachusetts in the seventeenth century. Some girls start off by lying about another girl but, as the deceit progresses and they become more emotionally invested in it, they appear to come to believe in the reality of the deception. These histrionic elaborations of emotion now seem fake to us, but in the emotionally unaware landscape of the seventeenth century, and in communities that inhabit emotionally repressive worlds, latent unexpressed emotions will find an out in some socially permissible way. Salem was a cruel and unhappy community, brimming with unspoken lies, and ripe for mass hysteria. In the end, the lie seems to become part of a collective belief because it fits with the emotional catharsis required by the group.

False Memory

Moving into the non-conscious territory of memory infidelity, we fall into the rabbit hole of 'false memory'. False memory is generally understood to be remembering events differently from how they happened, or remembering things that may not have occurred at all. The main problem with this framework is that 'true' event memory is a contradiction in terms. The adjective 'true' is usually not used for memory, with good reason, but there seems to be a general acceptance that events can be recalled as they originally occurred, in spite of what science has described since the nineteenth century. Memory, as Bentley proved in 1899, is not reproducible as a stream of sensory experience. All biographical memory is false to some degree, because of the imperative of change, the changing networks caused by ongoing events and experience, and the human drive to self-narrativize.

One of the most loved writers, Nobel Laureate in Literature Alice Munro, put this so well: 'Memory is the way we keep telling ourselves our stories – and telling other people a somewhat different version of our stories.' As she says, we first tell ourselves *our [own] stories*. Then we polish it up a bit, or a lot, and tell 'other people a somewhat different version', and the different version becomes the new version that we 'keep telling ourselves' and subsequently change again for others. Self-narrativization transforms your story into your narrative, and is what you want yourself to be and how you want others to see you. It is all, ultimately, driven by vanity. One of the most striking and attractive features of a humble person is their lack of a need to self-narrativize. The world has been exposed to some spectacular narcissists, whose vanity and sense of self-importance is equalled only by their deluded self-narrativization. We can hide behind our narratives, we can place ourselves complacently, sometimes almost comically, in the world as a made-up person. An individual's capacity to erase, reconstruct, selectively forget, selectively remember, is endless. So, if past memory, whether it be of a card colour, or of autobiographical events, is being constantly reconstructed by the demands of current experience and our changing self-narratives, is there any such thing as 'false' memory?

What Are We Talking About When We Talk About False Memory?

There are a tangle of ideas and terms related to the phrase 'false memory'. Psychiatry was in the thick of speculations about repressed/suppressed memory in the era of psychiatry dominated by Freudian thinking for most of the twentieth century and which only ended when brain science began to come of age in the latter decades of the century. The concept of repressed as opposed to suppressed memory was taught to us in earnest – suppressed memory was when a person deliberately pushed a memory from consciousness, and repressed memory was when this was not done deliberately, that is, when it was an unconscious process. I have been at case conferences – where individual patient cases are discussed among colleagues – and the issues of fabrication versus suppression versus repression have been discussed *ad nauseam*. Was the patient 'genuinely' totally amnesic? Was the patient feigning memory loss for secondary gain, perhaps because he didn't want to return home and face his gambling debts? The case conference discussions were, inevitably, circular. The usual course of psychogenic amnesia described in case studies in the psychiatric literature is that the amnesia resolves gradually if the treating team enables memory function to return without loss of face. Exploring the reasons, the dirty secrets of Salem, the domestic conflicts, the secret addictions, the avoidance, is more fruitful than exploring whether the story is true or false. For the record, I have never seen a case of so-called 'psychogenic' amnesia.

Freud is the inventor of the befuddling concepts of repressed and suppressed memory, and the subsequent confusions about false memory. This was based on his ideas about child sexual abuse. Freud moved from a theory that female neuroses – an umbrella term used to describe mental ill health – were caused by childhood sexual abuse to the theory that this was fantasy. He theorized that 'infantile sexuality' led to girl children being attracted to their fathers. Childhood sexual abuse, he went on to theorize in increasingly obscure language, was not an event that had occurred but a fantasy in the girl's mind. In 1933 Freud wrote about 'memory being transformed into fantasy', fantasy being the sexual abuse.[ii] Was Freud aware that he had borrowed Bentley's phrase about memory passing into fantasy, or was his action unconscious, repressed or suppressed? The societal denial of incest and child sexual abuse fell away in the closing decades of the twentieth century, and it became apparent that childhood sexual abuse was sadly common. Then something quite bizarre occurred, and, in a strange contortion of Freudian theory, Freudian techniques of hypnosis and suggestion were used to 'recover' memories of childhood abuse. The practice of recovering memories of childhood sexual abuse using Freudian techniques of stimulating repressed memory then developed into another monster.[137] Memories of past abuse were being induced by psychotherapeutic techniques of suggestion, by encouraging patients who had 'blocked' memories to attend survivors groups and try to recover the memory, or by giving them suggestive books.

A debate about whether recovered memories were 'false' or 'true' followed. An argument about a recovered versus a false versus a true memory might as well be about how many angels can dance on the head of a pin. Elizabeth Loftus has been an important figure in bringing some reality to the area of false memory and, importantly, in removing techniques of memory manipulation from mainstream clinical practice.[138] So-called recovered memory is no longer allowed as evidence in court in most jurisdictions. It is important, and often left unsaid, that unfortunately the victims and survivors of child sexual abuse do not need memory stimulation techniques to recall the ugly events that poisoned their childhood, their foundational memories and their mental health for the rest of their lives.

The Neuroscience of 'False' Memories

Although psychiatry has abandoned the idea of false memory, it persists as a term in the neuroscience literature. Given the disservice that was done to the countless 'hysterical' women who were victims/survivors of sexual abuse in the post-Freudian era, I feel that it is regretful that the adjective 'false' should precede 'memory' in any discipline but, adjective aside, the neuroscience of false memory is a fascinating coming together of discoveries with potentials that are far-reaching for human memory systems. The new neuroscience of false memory started in the biology of algae, something that is a familiar sight in my life.

I live in a village called Howth, on an island joined to the mainland only by a road, on the northern tip of Dublin Bay. The road to Howth from the city hugs the coastline, passing miles of sea. We have had so much sun in recent summers that when I drive or cycle home from the city, and when the tide is out, the sea is transformed into a massive field of a neon-green algal explosion. On first sighting the fluorescent green left glowing on the strand, I experience a memory tug of an amazing scientific experiment by Susumu Tonegawa. Tonegawa is a scientist who crosses multiple disciplines and who has won a Nobel Prize for his work in immunology.[iii] His story starts with the development of a techniques called *optogenetics*, which was called the great discovery of the twenty-first century.[139]

To understand the rudiments of the extraordinary scientific story of optogenetics, we need to start with the neon-green alga. The green colour comes from pigmented proteins, called rhodopsins, on the surface of the algal cells. The rhodopsin protein is really a channel that allows sunlight to enter the cell, and this light energy is then transformed into cellular energy that allows the alga to move and divide. This process is similar to what chlorophyll does in leaves – transforming light into cellular energy – and to what retinal pigment does in the eyes – transforming light energy into electrical neurotransmission. The human retina produces rhodopsin as one of several light-sensitive proteins. In all of these examples, light energy is converted to cellular energy through pigmented molecules that bring the light energy into the cell, be it a green rhodopsin molecule in an alga, a green chlorophyll molecule in a leaf, or a red, blue or green retinal cell. This is the basis for one half of the new science of optogenetics. The other half of the story involves genetic engineering, manipulating human cells to produce the rhodopsin protein, hence the name *optogenetics*.

How was this achieved? As evolution would have it, rhodopsin is produced in retinal cells, as we've already noted. The protein is only produced in the eye, but the rhodopsin DNA is present in a latent form in all cells: the DNA is present in brain cells but it is not made into proteins except in the eye. A scientist in Detroit, Zhuo-Hua Pan, had been working for years on trying to insert rhodopsins into non-functioning retinal cells in blind mice. He eventually succeeded in doing so using a microbe that carried the genetic code for the rhodopsin protein.[140, 141] Blind Mouse began to produce the rhodopsin protein, allowing light to activate the retina and giving Mouse sight. Simultaneously, collaborating scientists from the US and Germany published a study of how hippocampal neurons in the brain could be genetically manipulated to produce rhodopsin proteins and then be activated by shining a light on the neuron.[142] [iv] Optogenetics has allowed neuroscientists to activate an electron by shining a light on it. This genetic engineering allows scientists to see real-time memory at play at the level of the neuron.

This technique has been exploited by Tonegawa to look at memory formation and, more recently, what he has called 'false' memory. Tomás Ryan, who recently returned from Tonegawa's lab in Boston to our Trinity College Institute of Neurosciences, has provided us with a vivid first-hand account of an experiment that he worked on with Tonegawa.[143] In this mouse experiment they labelled the memory of an object – a bland and unremarkable blue box – by introducing rhodopsin into the hippocampal neurons that composed the cell-assembly memory of the blue box. Next, they put Mouse into a red box in which she was frightened by electrical shocks delivered through the floor of the box, and she exhibited the typical fear response of freezing. Mouse now had an emotionally neutral memory for the blue box and an emotionally charged memory for the red box. The next stage of the experiment involved putting Mouse back in the blue box while *simultaneously* activating the red box cell-assembly by light. When Mouse was then returned to the previously emotionally neutral blue box she froze with fear.

Fear had been transferred from the red box to the blue box memory. They had manipulated the blue box memory by tagging an emotion to it.

This brilliant experiment has just one problem for me as a psychiatrist – its title: 'Creating a false memory in the hippocampus'.[143 v] The title interested me because I had not considered that inserting an emotion into an emotionally neutral memory was creating a false memory, because this is what we do all the time. For example, a parent could be gentle and benign to a child almost all the time, except when they get drunk. The fear induced by the experience of a capriciously aggressive and out-of-control parent changes the memory of the parent for the child, because it is now tagged with emotion. The second issue that one could question is whether a memory that is made artificially is a false one. Regardless of how a memory is induced, whether through internal stimulation of Mouse neurons or through exteroceptive perception, the neural matter of the experience is formed. Is an auditory hallucination of voices, when voices are heard from outside the person, not as real an experience as a real person speaking? One can say the event did not happen in a shared common reality, but the experience of either an auditory hallucination or of a real person speaking are both grounded in the neural matter of memory. What one can say definitively from the Tonegawa experiments is that memory can be modified by artificial means, and this is far more fascinating than whether the modified memory is true or false.

There are many potential clinical applications for optogenetic techniques in treating human brain disorders, although this is some way away as yet. An optogenetic technique could hypothetically be used to make a brain implant that could be activated with a light, like a light-activated pacemaker, to modify brain function. Optogenetic techniques can now also be used to turn off neural activity. A study conducted in the Institute of Neurology at University College London has used optogenetic techniques to successfully control epileptic seizures in a rat cortex.[144] Epilepsy is due to excessive indiscriminate firing of cells, and in this study inhibitory neurons were genetically modified to fire when exposed to light, thereby inhibiting firing in the damaged rat cortex.[145] Medical science has gone from a surgical resection of HM's hippocampus for the treatment of intractable epilepsy in 1953, to light-controlled inhibition of seizure activity in 2012 . . . in only fifty-nine years.

Another important research application for optogenetics is using the technique to see what specific parts of the memory circuitry may be affected by particular treatments.[146] Will optogenetic stimulation of damaged hippocampal neurons allow repair and regeneration in depression and dementias?[147] Will it be possible one day to attend a brain clinic to regenerate in a targeted way damaged memory circuitry? Trauma, perhaps? The optogenetic story shows how the bringing together of disparate expertise – genetics, physics, natural sciences, neuroscience, optics, medicine, bioengineering – can lead to phenomenal leaps in applied knowledge. In the new connected-up era of the worldwide web and inter-disciplinary scientific collaborations, this can now happen.

New Ideas on Balanced Inhibition of Memory

Of relevance to memory retrieval is the new science of exploring what we remember and what we forget. Sensory input is always pushing its way through relatively stable tangled synaptic organization. The dynamic between the consolidated existing structures and the disruption caused by current experience is now being explored and is called 'competitive maintenance'.[148] I have been given insights into the dynamics of memory excitation and inhibition by my colleague and friend in our neuroscience institute, Mani Ramaswami, who investigates the molecular mechanisms of this dynamic through his work on drosophila, the common fruit fly, whom I no longer swat.[149] The lesson that I have learned from Mani's work is that neural inhibition is an active and necessary process in filtering out sensory information, without which there would be chaotic sensory overload. I also find it fascinating that the complex process of selectively attending to and remembering can be measured at the level of brain cells, and even molecules. What can be intuited through self-awareness and introspection can be explored and measured psychologically and can also be gleaned through measuring behaviour. At the same time, it is always occurring at a cellular level, even if what emerges is a sum of millions of positive and negative charges.

Attempting to answer any one question in neuroscience research seems to throw up more and more questions, often opening up whole systems of high complexity that have not been explored at all. In the process of examining what is memorized, a new process of inhibitory balance is opened up: there are so many possibilities in neural physiology, a staggering possibility of outcomes in neural connectivity and network formation! All we really know is that there is an immediate sensate world that enters the brain through sensory pathways carrying signals that jostle within synaptic tangles of inhibition and disinhibition, within circuits, within networks, and which may lead to temporary or permanent changes in neural architecture. Although the individual neural processes are ultimately determined by laws of science, even if we understood these laws the range of outcomes would always be infinite. How can memory ever be true to events?

Lost Domains

Tonegawa believes that one day it may be possible, through optogenetic stimulation of the amygdala, to reactivate happy memories. Neuroscience is now encompassing the neural dynamics of time, but Tonegawa's dream of finding Proust's lost time of past happy memories is, I think, an illusion.

Is there ever a boundaried memory untouched by the present, like a walled cement garden? It may feel as if there is something that lingers almost undisturbed in personal memory: a place that exists but which cannot be re-entered, a place that we intuit, a 'vanished loveliness', a lost domain.[vi] One of the great classics of French literature – written in the early twentieth century, in the same decade as Proust's *In Search of Lost Time* – is *Le Grand Meaulnes* (*The Great Meaulnes*). Also translated as *The Lost Domain*, it is a search for a mysterious lost world in the woods. The domain is remembered with a dream-like consciousness, in which vivid images shift with little sense of temporal connection, and childhood memories merge with the beginnings of sexual awakenings. The dream-like images and feelings that Meaulnes recalls as being located somewhere in a maze of woods and fields seem to me to be in the forests of the arborized roots of dendritic connectivity . . . remembered as a vanished loveliness. Franz, the narrator, tries to warn Meaulnes that you cannot travel back to the past but the hero fails to realize this, with tragic consequences. Searches for lost times and lost domains can never yield the same experience, however reluctant young hopeless romantics may be to accept this.

15. The Oldest Memories

Is the past ever really past? . . . ancestral zones are not contemplated as relics of a past that is over, but are rather intuited as persistent energies animating events, relations, and subjectivities in the present.

Anne Mulhall*

Collective Memory

It is obvious that it is from others that the ideas and contexts within which we understand the world are learned. Collective memory is usually conceptualized as cultural memory, but much of our deep collective memory is also biological. Although we are born as a relatively blank slate to be imprinted by experience, humans are first and foremost the biological product of living things that preceded them in evolution. All life forms on Earth share a genetic collective. One's unique genome is composed of the genetic material not only of one's family but of the collective genetic pool of our genetic ancestors, from algae to apes.[i] It is the organization of the cells in an organism that determines the sort of life that will emerge from clusters of cells.

Brown Bear and Matrilinear Memory

Paula Meehan, a much-treasured Irish poet, and an alumna of Trinity College Dublin, has written a poem, 'The Solace of Artemis',

* Anne Mulhall, 'Memory, Poetry, and Recovery: Paula Meehan's Transformational Aesthetics', in *An Sionnach: A Journal of Literature, Culture, and the Arts*, Vol. 5, Nos. 1 and 2 (spring and autumn 2009), p. 206.

that explores deep biological memory. The poem was inspired by an academic paper authored by a research collaboration between TCD, Oxford and Penn State universities, and concerns an Irish brown bear.[150] Brown bears have been extinct in Ireland since the ice age, some 20,000 to 50,000 years ago. The remains of one of these brown bears, a female, were discovered in 1997 in a cave in the Sligo mountains, in the west of Ireland. Finding Brown Bear's DNA in Sligo was exciting, but what was even more exciting was that a particular part of Irish Brown Bear's DNA, mitochondrial DNA, is present in *all* polar bears in the Arctic. Mitochondria are present in all cells and make the energy for the cell. They are often called the powerhouses of the cells, and are like independent little organelles within the cell that have their own DNA. Mitochondrial DNA code is inherited, unmodified, from the mother, in most species, including humans. This is called matrilineal inheritance. This means that Irish brown bears were the ancestors of all polar bears. Somehow, one or several female Irish brown bears made their way to the Arctic – the land masses of the Arctic and Europe were connected then – and mated with a native Arctic bear. Mitochondrial DNA, spreading unmodified from the west of Ireland for more than 40,000 years, is still firing cells with living fuel in all female polar bears. Paula contemplates the eternal deeply embedded biological female memory, comparing it to the superficial memories of the 'children of the machine', the first internet generation:

> their talk of memory, of buying it, of buying it cheap, but I,
> memory keeper by trade, scan time coded in the golden hive mind
> of eternity. I burn my books, I burn my whole archive:
> a blaze that sears, synapses flaring cell to cell where
>
> memory sleeps in the waxed hexagonals of my doomed and
> melting comb.*

* Paula Meehan, 'The Solace of Artemis', from *Imaginary Bonnets with Real Bees in Them* (University College Dublin Press, 2016), p. 30.

Something persists in the depths of the eternal female memory, perhaps romantic and maternal love . . . *waiting in the cave mouth for her lover loping across the vast ice field . . . 'dreaming my cubs about the den, my honied ones, smelling of snow and sweet oblivion.'*

In a further twist, mitochondria themselves were probably once bacteria that became incorporated into animal cells and became subverted to the purpose of survival of the cell that engulfed it.[151] The mutated bacteria continued to have semi-independent life, nestled in the cellular machinery of every human cell. Viruses have identical structure to RNA, the code assembly in humans that translates DNA into proteins. Viruses, too, are interlopers that fall into the human cell's DNA machinery, and they would not have any effect on us if we did not share the same molecular biology. It is because COVID-19 can trick human cells into producing the COVID proteins that humans become diseased. This can sometimes be put to our advantage, and viruses are now used as carriers to change DNA production in living human cells that are diseased. We also embody our evolutionary predecessors in memory and emotional experiences. For example, as we've explored, humans are wired to immediately respond emotionally to smell memory. This is because older evolutionary species, in the absence of conscious reflection, needed to recognize and respond immediately and automatically to danger. The short-circuit from smell to response keeps Rat, who has a relatively huge olfactory cortex compared to us, alive; and we are the lucky inheritors of this emotional experience. Place probably has a central role in the human memory system because of the importance for predators and food gatherers of returning to places where food is plentiful and avoiding places that might be dangerous. A lot of hard-wiring in the human brain is present, because of the demands not of the human world, but that of phylogenetic predecessors.

Cultural Memory

While deep biological memory is humming away in the background, cultural memory is to the fore in the way we construct new memories and put it all together in order to understand the world. Paul Baltes and Tania Singer from the Max Planck Institute in Berlin have summarized the inseparable contributions of biological and cultural memory to individual memory systems:

> There is general agreement that the mind is the bio-cultural co-construction of two interactive systems of influences: internal genetic-biological and external material-social-cultural. The brain is a joint outcome of these two systems of inheritance.[ii]

While recognizing the inseparability of biological and cultural memory, they conclude that socio-cultural influences are more dominant in the modern world. What I see and perceive leads me to believe that, in an average life, nothing is more important to what we attend to and memorize than our osmotic relationship with socio-cultural memory.

The term collective memory, *mémoire collective*, was first coined in 1925 by the French sociologist Maurice Halbwachs. Halbwachs' core thesis was that an individual's private memory existed within the framework of collective memory, without which private memory could have no meaning and no context: 'Yet it is in society that people normally acquire their memories. It is also in society that they recall, recognize, and localize their memories.'* He believed that one set of cultural beliefs or memories, which he called 'frameworks', are gradually colonized by others as societies change. The gradual introduction of a new set of ideas into a pre-existing framework means that the whole can maintain a stability

---

* Maurice Halbwachs, *On Collective Memory*, ed. and trans. Lewis A. Coser (University of Chicago Press, 1992), p. 38.

while constituent ideas gradually transmute. One can see this in a simple form in the preservation, and renaming, of rituals from pre-Christian to Christian times, the prototypic example being the takeover of the pre-Christian celebration of the winter solstice by the Christian Christmas.

Perhaps you are noticing that the dynamics of cultural memory are similar to those of individual memory – not fixed or unchanging, but in a process of being continually reconstructed by the present. It is as if there is a collective human cortical web of organization, sometimes disorganization, that is loosely consolidated but is being modified constantly by vast streams of sensory input. As Halbwachs observed, the past is not preserved but is reconstructed based on present beliefs. His description of this process of revisionism is the same as that for individual memory, where a disruption of memory occurs because of shifting self-narrativization: '. . . we choose among the store of recollections . . . to an order conforming with our ideas of the moment.' The fragile stores of collective memory seem to be just as mutable as those of personal memory.

The Oldest Stories

One constant, however, in the ever-shifting heave of human culture is the fairy tale. They are the oldest stories, universal but woven into the local, and handed down through the spoken word from generation to generation. I was reared in the culture of the local fairy stories because, surprising as it may seem, my generation in rural Ireland were still very close to the tradition of transmission of stories by the spoken word. I was, and remain, embedded at a deep level of memory in the power of the fairy story. It took me some years to recognize the hidden motif of postpartum psychosis in fairy stories. You were probably jolted, as I was, not only by Edith's terrifying psychotic delusions but also because it was redolent of culturally familiar stories of the possessed infant/child. Edith's experience had a more distant sense of familiarity for me than the booming resonance of Hollywood movies, more like a grumbling undertow.

Irish fairy stories, similar to those elsewhere in Europe, were generally only committed to paper at the turn of the twentieth century. At this time, a movement developed in Ireland, the Celtic Twilight – similar to that led by the Grimm Brothers in Germany a few decades previously – to preserve indigenous national culture.[iii] The penning of Irish folklore had a special significance because Irish language, history and religion had been outlawed for several hundred years during British rule. Hiberno-English was spoken in most of the country, which was a fusion of the grammatical structure of the Irish language with English words. Spoken Irish did, and does, remain the main language in small areas on the western, northern and southern seaboard, where the fairy stories were most likely to be told. The laconic Irish language, with its relatively fewer words and subsequent greater ambiguity, is a perfect language for the pithy fairy tale.

Douglas Hyde provided the first unadulterated written account of Irish fairy tales and folklore in 1890, using the 'exact language' of the people (W. B. Yeats's description). Hyde was a leader in the illuminati of the Celtic Twilight, some of whom, neither Yeats nor Hyde though, went on to become revolutionaries in the war for independence that started with the 1916 Rising. Hyde collected the stories as he heard them, in Irish, and wrote an English translation on the facing page.* I have, open beside me, an original edition of *Beside the Fire* that belongs to my neighbour Mary, who was married to Douglas Hyde's grandson. Douglas Hyde, the passionate penner of the bilingual Irish fairy story, went on to become the first president of the new Irish republic in 1938. Small surprise, then, that fairy stories were so interwoven into my young imagination.[iv]

My siblings and I grew up steeped in the magic of Irish fairy tales. We spent our summers on the small farm where my mother's family

---

* Douglas Hyde, *Beside the Fire: A Collection of Irish Gaelic Folk Stories* (David Nutt, 1890).

had lived for several generations in County Kerry, in the southwest corner of Ireland. We moved house several times when growing up, but summer holidays in Rathoran were a constant. The prefix 'Rath' – common to many Irish place names – is the Irish/Gaeilge for a raised, circular area of land with a flat top rimmed with trees, usually haw-thorns, that were considered to possess magical powers. It was said that fairies lived under the raths. Uncle Jim – Mother's brother, who ran the farm with his sister Aunty May – told us that a neighbour who was trying to dig up a hawthorn had his tractor repeatedly pushed back by the hawthorn tree. Uncle Jim himself did not have a tractor and continued to rely on a horse and cart for the transport of the milk pails to the creamery each morning, until the horse died. Some of my most precious memories of these summers are of getting up early and sneaking quietly out of the bedroom so that I would be the only sib-ling to go to the creamery, being lifted on to the back of the cart by Uncle Jim, my legs rattling to the bumps on the dirt road that led from the farm to the main road to Abbeyfeale.

'Simply Life'

W. B. Yeats also fell under the spell of the Irish fairy story, and in the Introduction to his book *Irish Folk Stories and Fairy Tales* he describes Douglas Hyde's work as 'neither humorous nor mourn-ful: it is simply life'. Yeats's well-known short poem 'The Stolen Child' is probably related to the common occurrence of death from unknown causes in children:

> Come away, O human child!
> To the waters and the wild
> With a faery, hand in hand.
> For the world's more full of weeping than you can understand.

This brings us back to Edith and her psychotic belief that her baby had been substituted by an identical double. This motif of the changeling child is universal. The changeling was believed to have been left by the fairies, who stole the real child. In one version of the changeling fairy tale, set down by Yeats, a woman, Mrs Sullivan, believes that her baby has been exchanged in a 'fairies' theft'. As is typical, she takes advice from a local wise woman. The wise woman advises her to boil egg shells in a big pot on the fire. Boiling shells would have been considered absurd – one would boil eggs, but never shells. If the baby is an innocent without any knowledge, they would not be surprised by the boiling of egg shells. If they were a cunning changeling, it was reasoned, they would be prompted to respond to the tomfoolery of boiling eggshells. As Mrs Sullivan boils the eggshells, the baby/changeling asks, '. . . in the voice of a very old man, "What are you doing, Mammy . . . what are you brewing, Mammy?"' The gift of speech 'now proved beyond question that he was a fairy substitute'. Mrs Sullivan's talking changeling is redolent of the auditory hallucinations that women with postpartum psychosis experience and frequently attribute to a substitute baby. It is Mrs Sullivan, not the wise woman, who then kills the baby – an outcome that is common in untreated postpartum psychosis.

What in folklore was known as a changeling is probably, in the nomenclature of psychiatry, a delusion of substitution.[v] The delusions of substitution in postpartum psychosis can involve everyone around the woman but most commonly her baby, her husband/ partner or, if hospitalized, the medical staff. When working in the Bethlem I became so accustomed to delusions of substitution that I would approach the idea in a gingerly way if women were in a shut-down state, reassuring them that their possible suspicions of substitution were part of the illness. One woman memorably believed that her husband had been substituted by the devil on the night that she conceived her baby. He was now her husband again, but for the night of the conception the devil had taken him over.

In about half the cases of postpartum psychosis, a woman has never previously had a psychotic episode and it strikes like a bolt of lightning from the blue. It is a bewildering experience for her, and for those close to her. In fairy stories dramatic changes in women who have given birth are ascribed to substituting the mother with an identical double. A fairy tale describing such a substitution was among those recorded by the Grimm Brothers:

> When the queen gave birth to a handsome prince the next year and the king was out hunting, the witch appeared in the form of a chambermaid and entered the room where the queen was recovering from the birth . . . The witch led the queen into the bathroom and locked the door behind her. Inside there was a brutally hot fire, and the beautiful queen was suffocated to death . . . Now the witch had a daughter of her own, and she endowed her with the outward shape of the queen and laid her in bed in place of the queen.

In other versions, the recently delivered mother is stolen by the fairies to suckle the fairy infants. When returned, possibly years later, she would often be greatly changed. Another motif is that of a woman who has given birth and becomes sleepless, and who eventually closes her eyes in exhaustion, whereupon a changeling is laid in the crib by the waiting fairies. This may reflect the sleeplessness of psychosis followed by delusions of substitution. The Fairy Doctor, the Irish version of the wise oracular woman, is another important aspect of fairy lore, and the Fairy Doctor often gave advice around birth. Writing about the Fairy Doctor in a small county in Ireland, the folklorist Erin Kraus has written:

> The in-between world inhabited by fairies could be glimpsed at the boundary of states of time – at dusk or at midnight, on Hallowe'en or May Eve; and the boundary of states of place – at the town's edge, between tides, at the garden's end . . . Likewise, transition states, such as birth, sexual maturation, and death, were associated with access to the fairy world.[vi]

The traditional spoken word of the local fairy story has morphed into universal narratives of Walt Disney wholesomeness, in which the Good prevail, the Innocent are rewarded, the Inept and Lazy have their comeuppance, and the Evil ones perish or are punished. These derivations do no justice to the largely unreconstructed stories of life, particularly female life, that were the subject matter of the unsanitized original versions. Marina Warner, a scholar of fairy stories, in her intriguing book *Once Upon a Time: A Short History of Fairy Tale* (2014) recounts the journey of the Brothers Grimm to a hospital in the German university town of Marburg, where they had travelled to meet an old woman, famous for her repertoire of folklore. The old woman refused to give the Brothers Grimm her stories. The brothers eventually acquired her story – that of Cinderella – through hoodwinking a young girl to get the old woman to tell her the story. The old woman of Marburg told the girl child and the Grimms then took the story of Cinderella down from the girl.

Warner writes, 'perhaps the old woman did not want the elite young men to see into the secret thoughts and dreams of revenge' that women experience?* In the original version of Cinderella, the 'ugly' sisters self-mutilate, cutting off their heels and toes to fit the perfect shoe of the hypothetically perfect female foot. Doves, the bird of romantic love and hope, peck out the eyes of the sisters through the vengeful spirit of Cinderella's dead mother. The subject matter of the original fairy stories is usually brutal, but it is difficult to feel anything because there are no subjective emotions written into the texts. As Yeats put it, they are neither *humorous or mournful*. One reads about children being expelled from home by filicidal parents, grandmothers being eaten by wild wolves because the wolf wants to eat the granddaughter, girls camouflaging themselves as goats to avoid being raped by their fathers, babies and children being kidnapped and held imprisoned for years, children being eaten . . . and yet there is no emotion in the telling of the story, and we feel little emotion when we read the original versions. The story moves from one event to the next as a child would tell a story.

* Marina Warner, *Once Upon a Time: A Short History of the Fairy Tale* (OUP, 2014).

The stories were transmitted in this way I think because the information that they contained, transmitted from woman to girl, was too important to be distorted by subjective feeling. It is not the interpretation of the story that matters, or how one may feel, or whether something is right or wrong; it is the fact that incest and rape and murder and sexual rivalry exist and the girl needs to know this and learn to protect herself. The stories were too valuable for life and survival to be transformed by narrativization. The power of the female transmission of these stories, the raw earthiness of the non-idealized objectified female, is for me the real deep female collective memory, the dark side of the eternal Brown Bear as mother and lover.

Some Concluding Thoughts

I have lived through a period in medicine that has gone from medical students being taught about the brain in terms of separate, functionally defined pathways with psychiatry 'located' somewhere in the black hole of the emotional-memory circuit, to the beginnings of the scientific understanding of the connected brain. There were no neuroscience departments in the 1980s and few in the 1990s – all this brain knowledge is new and has happened in the blink of an eye in historical terms. I have learned about the brain as this knowledge has unfolded, but my reference points for this new knowledge, my foundational memories, are rooted in personal experiences, in the experiences of my patients and in the great creative thinkers who intuited what we are now coming to understand scientifically – great artists who, immersed in introspection, wrote about the experience of memory before we could name the processes involved. I have learned, as we all do, through both knowledge and experience. Foundational memories may be based in established scientific knowledge or in the collective wisdom of the unmodified fairy tale, or the genius of highly creative observers.

The merging of knowledge and experience can happen because we have a multi-tiered brain memory system, where new experience is held in the plastic hippocampus, the memory maker, and gradually integrated into the less plastic, more consolidated cortex. Current experience and memory are integrated in the complicated networks of the prefrontal cortex, the storyteller. At the highest pinnacle of this complexity, memory is consciously manipulated in imagination. At this level, memory can be worked without external sensory input, and this faculty can be used to form new patterns of thinking, to imagine and create, to modify one's understanding of the world. It is through this representative memory that we develop self-awareness, and the appreciation of others as being self-aware in the same way. Through this we come to accept the singular human state, common to everyone, of existential aloneness and inseparability from others. The appreciation of others as being mirrored humans is the neural basis for the virtue of human kindness.

The core expertise in psychiatry is in understanding experience and naming it. That is why individuals with psychiatric illnesses have a great deal to tell neuroscience, and the larger world, about the processes involved in the organization of memory. Psychiatric illnesses probably involve disruptions in integrative brain processes, in network brain function, which we are only coming to understand now through network neuroscience and connectomics. As this science develops, psychiatric illnesses will become a major target of investigation, and I believe this will be the beginning of the ending of the stigmatization of psychiatric illness. An important caveat here is that most of my patients do not feel similarly optimistic. They are not immune to the stigma of mental illness, because they suffer from it, and, even if they have overcome their own internalized stigma, they know that, in general, others have not. Some pain is inevitable in life, from sickness, separation, death, failure, and some is unnecessary – perhaps this is the most painful of all suffering.

Back in the days of the Bethlem woods and the badger sett, what

I really learned, although I didn't know it until years later, was that Edith's flashback, a memory of an event, was also an experiential event. Her experience on her first sighting of the small gravestone had been translated into a specific neuronal imprint that coded for the gravestone and the associated terror. The re-sighting of the gravestone the second time ignited the arborization of neurons like a flash of lightning that lights up a tree in a storm, creating a fresh experience. Edith brought home to me how memory is, in essence, neurally coded experience. This experience can be reactivated like a dramatic reenactment and cause emotional distress, or it may tweak a subterranean intuition from the buried rag and bone shop of the past.

Neuroscientists may talk about memory as a process, psychiatrists may talk about it as a repository of experience, and neurologists may talk about pathology in specific brain areas leading to specific deficits in memory function. But memory derives from the infinite jostling and flarings of neurons in a vast, pervasive connectivity of present experience interacting with memory networks. All the microscopic synaptic activity flashing on and off like fairy lights has a whole-brain effect and it is the whole-brain effect that presents itself as a conscious experience. In Edith's words 'the memory was *real*', an irrefutable experience. Ideas come and go, adrift on a sea of cultural zeitgeists, but living human experience, at the end of the day, is bigger than ideas – just like the brain, experience is irreducible.

Postscript

At the end of every scientific article the authors write a section, titled 'Limitations', about how their study is flawed. Small details are listed explaining why the findings may not be correct or generalizable, and the authors usually conclude by saying that the study needs to be replicated, addressing the stated limitation of their work, before the findings can be confirmed. This ambiguity and hesitation is the opposite of what the world outside of science desires. There is a general desire for neat, simple information and a general intolerance of ambiguity. Medics have to embrace ambiguity and often take an educated guess based on experience, and will continue to do so until we have machines that can precisely evaluate the various stages of function or failure in body systems, the causes of the failures, *and* we have developed biological or pharmacological therapies specifically targeting the pathology. Psychiatry will be last to arrive at this point, because it deals with the most complicated aspects of the most complicated organ. So I will end on a note of caution by emphasizing that the contents of this book reflect the state of current knowledge that is advancing as I write, and you read, the book.

I have a headful of anticipated events as I emerge from this lovely summer of 2018. I will not – for as long as I will remember – forget this long hot summer heralded by the joy of the Irish referendum. The sea shadowed by the cliff in our north-facing swimming bay in Howth was spared the algae bloom at the cost of the chill. Following a swim to the far buoy I would often spread myself out on the stones, warmed by the heat, and feel the prescient memories being fired by whatever it was that made the happiness come together that summer. I swam in the sea and I swam in memory.

The warm stones, the late evening swims in mad rose-orange sunsets and the early morning ones of still luminescence, the swimming community chatting during the ritual of drying and dressing, no detail of the swim too small to elaborate upon, the heightened awareness of the comings-and-goings of my eldest's last summer at home, and the daily writing. I must now leave these immersions and face the shortening days. Gaston Bachelard called winter 'the oldest of the seasons . . . confer[ring] age on our memories, taking us back to a remote past'.

Thank you for reading the book, or parts of it, and I hope that you have enjoyed reading it with even a small fraction of the pleasure that I had in the writing of it.

Thank you . . .

Esther and Sean, mother and father, and siblings, Joe, Myra and Therese, for the memories. A special thanks to Joe, who read one of the original drafts and gave detailed feedback.

Ted (Dinan), my first mentor and a friend who allowed me to learn by paring down concepts that were obtuse to piecemeal information that I could reconstruct.

Robin (Murray), my second mentor, who demonstrated to me how to follow your own questions.

My patients, all of them. I feel an enormous sense of gratitude to those who have generously agreed to let me share anonymized extracts from their case histories for others to learn from.

The scientists who have given the world so much clarity about how it all works. I have mentioned only some of my favourites, a small fraction of the many great scientists. A big thanks to my neuroscience colleagues in the unique Trinity College Institute of Neuroscience, where psychiatrists, neurologists and laboratory scientists work side by side in an atmosphere of happy and excited knowledge-sharing. In particular I would like to thank the neuroscientist-author Shane O'Mara, who introduced me to the world of trade publishing.

My fellow psychiatrists, whose collegiate exchanges and *know-ingness* about our shared experiences gives us a unique bond.

Mary (Cosgrove), with whom I collaborated in a neuro-humanities project called *Melancholia and the Brain*. I hope that Mary learned half as much from me as I did from her during our conversations on 'calcified dichotomies'.

Paula (Meehan), the poet. We have a special place for the poet in Ireland, going back to Celtic times. Among the many illuminations that followed from conversations with Paula was an understanding of the calling of the poet to be an authentic memorialist of the times. Paula, like Proust, holds close to the centre: an observer of her own, and other people's, experiences.

The visual artist Cecily Brennan, whose creative insights have led me to places that I would not otherwise have gone.

Daire, for helping me organize the original illustrations.

Sorca (O'Farrell) for doing the illustrations, and her husband Chris (Cawley) for reading the manuscript.

My friends, for the laughs and fun, for sharing the ups and downs, the traumas and the memories.

Cian and Rowan, who have given me great joy and the loveliest memories.

Bill Hamilton, my wise and clever agent and part-editor, who saw the manuscript in its original unformed state and helped me to understand what I was writing about.

Josephine Greywoode, my editor, without whose sharp eye and ear there would have been little sense of the entirety of the book.

My swimming community, especially the Orcas 2, for the daily restorations to equanimity.

Notes

1. Dawnings

i. There are several traditional classifications of memory, but these fall into two main categories. One is based in *time*: short-, medium- and long-term memory. These distinctions are useful clinically when trying to grossly evaluate memory function: for example, short-term memory becomes more impaired as dementia advances, while there is relative preservation of long-term memory; in a very severe brain injury there may also be immediate loss of long-term memory. The other major classification uses *type* of memory and is generally divided into implicit (also known as 'non-declarative') and explicit (also known as 'declarative', or 'biographical'). 'Implicit' relates to things that are automatic to us, for example motor function; while 'explicit' relates to conscious recall, for example event memory. I think that these divisions make it more difficult for non-specialists to understand the common processes involved in memory formation and recall, and for this reason I have not classified memory into separate categories.

2. Sensation

i. Charlotte Perkins Gilman wrote 'The Yellow Wallpaper' in 1892, although she suffered her mental breakdown during her pregnancy and following the birth of her daughter in 1885. She did not write again until 1890. She ascribed her illness to her unhappy marriage in 1884 and the subsequent birth of her daughter Katharine in 1885. The six missing years between 1884 and 1890 were a mystery that was given some explanation by the publication in 2005 of a letter that Charlotte wrote to Silas Weir Mitchell in 1887 requesting a consultation for treatment with the 'rest cure'. In this letter, which Charlotte kept secret until close to her death, she reveals the experiences of her lived psychotic depression.

Weir Mitchell was a prominent neurologist who treated neurasthenia – a catch-all diagnosis that included what we would now call post-traumatic stress disorder, depression, anxiety and bipolar disorder. He was fashionable in contemporary American society and treated well-known figures such as

Walt Whitman and Franklin D. Roosevelt. (Denise D. Knight, ' "All the Facts of the Case": Gilman's Lost Letter to Dr. S. Weir Mitchell', *American Literary Realism*, 37:3 (spring, 2005), pp. 259–77.)

ii. Copernicus (1473–1543) had ignited a scientific revolution with the explosive idea of a heliocentric model of the planets in the early sixteenth century. Galileo Galilei (1564–1642), born twenty-one years after Copernicus's death, expanded on this idea and was tried by the Roman Inquisition in 1615, found to be heretical and put under house arrest for the rest of his life. Copernicus and Galileo demonstrated laws of nature, and these were not compatible with God's Law. Exile and often death was the fate of many of the few who followed and explored a scientific explanation for the world.

iii. Autoimmune diseases are common and occur in most body tissues: joints (rheumatoid arthritis); thyroid gland (thyroiditis); gut (Crohn's disease); heart (cardiomyopathy), and so on. NMDA encephalitis, like many autoimmune diseases, responds to immunosuppressive therapies, and this may be the end of it, or it may recur again, as with most autoimmune diseases.

iv. The members and followers of these salons lived very 'alternative' lives, charged with passionate exploration of the new order that was emerging scientifically and politically. Women wrote, studied and argued alongside their male counterparts. One such salon, led by Madame Helvétius, was a much-coveted retreat for the intellectual outcasts who were becoming increasingly threatening to the *ancien régime*. She was the widow of the Claude-Adrien Helvétius, author of *De l'esprit* – a French compilation of ideas that combined sensationalism and sensibility – in which he argued for the equivalence of all minds, regardless of race or class and, more radically, of women with men. Following her expulsion from society and her husband's death, Anne Helvétius bought a small estate in Auteil, a suburb of Paris, and lived the life of a free woman. Some followers of her salon became permanent residents. Interestingly, Benjamin Franklin regularly visited her salon and seems to have fallen in love with her, before carrying back the French Enlightenment ideas to America. If you would like to read more about the lively debates that shook the world view, and the characters who populated these salons in France, I would recommend George Makari's book *Soul Machine: The Invention of the Modern Mind* (W. W. Norton, 2015).

v. This quote is from Sylvia Siros of Babylab, a unit set up in 2005 at the University of Manchester to look at cognitive development in infants. In Trinity College Dublin we are doing a longitudinal study of 100 babies born in Dublin in 2014–15 to women who were either depressed or not depressed during pregnancy, to look at the effects, if any, of depression during pregnancy on detailed measures of infant neurodevelopment.[74] This study is being done in collaboration with colleagues in the Department of Psychology, in their Infant and Child Laboratory. It is fascinating to see how the behaviour of the infant and the infant–parent interaction can be micro-analysed. Even glances are measured. A broad general finding is that childhood abuse or neglect in pregnant women is likely to lead to depression during pregnancy; if this resolves the infant has no developmental disadvantage, but if the depression is recurrent the infant is vulnerable to a poorer neurodevelopmental trajectory.[152] We have not yet completed testing our infant cohort at year 3.

3. *Making Sense*

i. MRI (magnetic resonance imaging) is an imaging tool used to measure volume of body structures. The MRI machine is basically a very powerful magnet that pulls some brain atoms in a particular spin, giving patterns of 3D structure. These patterns are read in comparison to standard brain anatomical maps of structure volumes. Functional MRI uses the same principle of magnetic stimulation of atomic particles to give patterns of blood flow in the brain over short periods of time. Blood theoretically flows to areas that are more active, and so will indicate that certain areas are being used when a particular activity is underway. In the Villringer study cited, the touch cortex had increased blood flow when the digits were moved, indicating that digit movement was controlled by certain regions of the touch, or sensory, cortex.

ii. Fiona Newell, a neuroscientist colleague, has shown how what might previously have been thought to be sensory learning in one domain, say sight, is actually spread around the different sensory cortices, both visual and auditory.[153]

4. *The Story of the Hippocampus*

i. Freud's theorizing, like that of other landmark thinkers, did not escape the sexist, genderist and racist prejudices of his era. Freud's hypotheses about the repressed universal envy of females towards male sexuality reflects the prevailing misogynistic views. Freud's sexualization of childrens' drives, however, does seem to go beyond commonly held beliefs of the era.

ii. http://www.richardwebster.net/freudandhysteria.html. This URL will take you to a succinct and comprehensive overview of the history of hysteria. Essentially, hysteria was a catch-all diagnosis for neurological and psychiatric disorders that could not be diagnosed. One of the most fascinating and famous studies looking at hysteria was that by Eliot Slater, in which he examined eighty-five middle-aged patients who had been given this diagnosis in the early to mid 1950s and followed them up over a period of nine years. Twelve patients had died, fourteen had become totally disabled and twelve had become partially disabled.[15] Most of these patients had been suffering from neurological disorders and had been misdiagnosed with hysteria. Slater wrote that 'The diagnosis of "hysteria" is a disguise for ignorance and a fertile source of clinical error.'

iii. One common term used for hysterical disorders, 'conversion disorder', derives from Freud's idea that emotional distress or conflict could be 'converted' into neurological symptoms, such as amnesia or paralysis. A diagnosis of 'conversion disorder' exists to this day in *DSM 5* and continues to be used in clinical practice. Sometimes, conversion disorders are called 'functional neurological' disorders. The use of the term 'functional disorder' supposes the occurrence of an event in the brain, such as a thought, as being separate from the brain matter that is involved in that function. It has been known since the early twentieth century that structure and function cannot be separated, even at a molecular level. This was demonstrated in humans by Christian Anfinsen in the 1950s, when he showed that a change in the structure of a protein transformed the protein chemically to do something different: structural change meant functional change. He won the Nobel Prize in Chemistry in 1972 for this discovery. 'Structure means function' is a basic principle of all sciences, including clinical neuroscience. For example, we interpret brain neuroimaging of structures as an indication of brain function, and fuzzy white coloured connecting paths are interpreted as representing poor axon formation, and hence poor connectivity. Even the size of a brain part is interpreted as being important – smaller areas imply poorer function. Dementia cannot be definitively diagnosed without a brain scan that shows atrophy of the brain. At a microscopic level, fewer receptors implies poorer function of their matched neurotransmitters, and so on. The division into neurological and functional persists in spite of all this knowledge.

iv. Molecular defects in memory formation in the hippocampus in depressed/ anxious models of mice are now being examined. In a Korean study published in 2019, the authors conducted some spectacular experiments in mice in which they changed a 'memory' gene in the mouse and this led to depression.[154] The memory gene is involved in synapse formation, and when you make it defective in mice, hippocampal growth is impaired. Not only did this group make a depressed mouse with a shrunken hippocampus, they went on to 'fix' the gene by giving the mice the protein that they were missing. This resulted in restoration of synapse formation and the curing of the depressed/anxious behaviours.

v. Biographical, or event, memory is 'stored' in the prefrontal cortex. It is not known whether the hippocampus is also involved in all event-memory recall. What is known is that the more vivid the memory, the more the involvement of other cortical areas – for example, if the recall is particularly vivid visually, then the visual cortex is more likely to be shooting signals through to the prefrontal cortex; if sounds or the emotion are part of the memory, then the auditory or the emotional cortex is more likely to be engaged; and so on.[155]

vi. *The Unnamable* is a stream of disjointed sentences in which the words create a feeling of existential terror because the person speaking does not seem to exist to themself. The narrator is lost, does not seem to exist except in the words, and is grappling to find some sense of being alive and coherent in the world. It seems like a fragmented state of mind brought about by trauma that has not been acknowledged by either the narrator or by the world. This sense of not being in the world is not dissimilar to some aspects of psychosis. The Beckett play *Not I* is largely drawn from *Unnamable*. Billie Whitelaw was individually coached by the demanding Beckett in her gripping and famous performance of *Not I*. This is easily accessed online.

5. The Sixth Sense

i. Willian Styron suffered from severe depression, probably bipolar disorder. His familiarity with extreme emotional states may have contributed to his exceptionally vivid descriptions of emotional experiences: '. . . casting a reluctant glance behind to the place where he had lain, and at the river and the cedars, too – disturbed about something: loveliness vanished, or perhaps merely the sense that one bright instant of his youth would always, mysteriously, be bound up in the invisible and fugitive smell of cedar trees.'

ii. Linda Buck and Richard Axel won the Nobel Prize in Medicine or Physiology in 2004 by demonstrating that each odour receptor, of which humans are estimated to have about 350, recognizes only one scent. Smell memory is much more important in mammals that are phylogenetically younger than humans, and the volume of brain matter dedicated to smell in rats is huge compared to humans. They remember through smell and therefore recognize and react secondary to smell. Mice have about a 1,000 odour receptors.

iii. Another theorist, Carl Lange, proposed a similar explanation for emotions around the same time (*On Emotions: A Psycho-Physiological Study*, 1885). Lange's theory, however, was less complex, proposing that the actual change in the body, the primary feeling, was the emotion; whereas James proposed that there was a secondary feeling generated in the brain from the primary feeling. Because the theories were similar in that both held that emotions were somatic, they were merged and it became known as the James–Lange theory of emotion. James's theory of emotion, over the years, came to be reduced to the less complex definition of Lange's: that emotion is the physiological changes, most of which are visceral. James, as we can now see, intuited that body feeling states were only part of the complex emotional feeling states of humans, and that body states would be modified by brain/memory responses.

iv. We do not have much control over autonomic or visceral responses, but individuals can, with various psychological techniques and intensive medative work, learn to modify or even control autonomic or visceral responses. In Daniel Goleman's *Destructive Emotions* (Bloomsbury, 2003) the author recounts neuroscience experiments with Oser, a Buddhist of more than thirty years. Oser had worked with some of the greatest teachers of Buddhism in Tibet and, in a collaborative exploration between Goleman and the Dalai Lama, agreed to be monitored by MRI as he was practising medatitive techniques. Oser, during one form of meditation, rather than having an increase in blood pressure and heart rate in response to being startled, had the reverse ANS response – a decrease in blood pressure and heart rate. But, unless we practice very hard, ANS and visceral responses are involuntary.

v. Increases in ANS arousal in the sympathetic system are accompanied by an increase in sweating. Having wet skin will cause a change in the ability of the skin to conduct an electrical current. This is the physiological basis for the electrodermal activity (EDA) test, used in lie detection. The electrodermal activity test conductance change can reflect feelings that one is trying to disguise. Most of us will have a transient negative feeling, a slight arousal, when telling a lie, unless one does not have feelings. One can consciously disguise feelings in facial expression, but one will get an increase in the automatically mediated sympathetic ANS activity, a component of which is sweating. This dampening will then alter the electrodermal activity test conductance. ANS arousal is also measured using heart-rate variability, respiratory rate and sometimes temperature. Those with a relatively unemotional personality, for example a psychopath, may not have these ANS changes and so do not have changes in their electrodermal activity test.

7. *Time and Experiencing Continuity*

i. There are rare instances in which a person may be awake, have a normal sleep–wake cycle but not be responsive to their environment.[156] Persistent vegetative or locked-in states are examples of these tragic human disorders of what is often called 'awareness' consciousness, the next order of consciousness up from wakefulness that we will explore in coming chapters.

ii. This is a quote from J. Clerk Maxwell's timeless gem *Matter and Motion* (1876). He was first a professor of natural philosophy, and was later appointed Professor of Physics and Mathematics in Cambridge University. He was a dazzlingly broad thinker who saw that the brain could not be left out of our understanding of the physical world, hence his observations about place and memory.

iii. It is worth clarifying Freud's complicated stance on childhood sexual abuse and memory. Up until 1897 Freud believed that patients, mostly women, developed neuroses or hysteria because they had been sexually abused during childhood. He elaborated his clinical observations in a theory called the 'seduction theory'. Although the name 'seduction theory' implies some sort of adult-like consent on the part of the child, the suggestion that fathers might be abusing their daughters went far beyond societal ideas at a time when incest was vehemently denied. The seduction theory was received with outrage by the medical establishment, and in 1897 Freud announced that he had been misguided about the cause of hysteria and neuroses in female patients and began to ascribe female neuroses to theories of masturbation and excessive menstrual bleeding; he even went on to consider the role of witchcraft. After abandoning seduction theory, Freud was committed to the idea that sexual abuse of girls was a fantasy. References for Freud's change of direction can be found at https://www.theatlantic.com/magazine/archive/1984/02/freud-and-the-seduction-theory/376313/.

iv. If you would like to explore this topic further I recommend Douwe Draaisma's book *The Nostalgia Factory: Memory, Time and Aging* (Yale University Press, 2013), in which he explores the increasing sense of the rapidity of time passing as one ages.

v. In Sean Carroll's book *The Big Picture: On the Origins of Life, Meaning and the Universe Itself* (Dutton, 2016), the author assumes that the natural world is the basis for everything. It is refreshing to read, because he invokes the brain in what may have been previously investigated by physicists. Quantum physics principles are now being incorporated into molecular biology, and I believe that we will move eventually towards an understanding of the quantum physics principles that underlie the behaviour of neurons.

vi. Two-directional place cells were first found in the hippocampus, then head-orientation cells brought memory to the three-dimensional level of space, and then time cells were discovered that integrated with space memory, giving us a four-dimensional timespace continuum.

'Henceforth space by itself, and time by itself, are doomed to fade away into mere shadows, and only a kind of union of the two will preserve an independent reality.' With these lines Hermann Minkowski famously opened his ground-breaking lecture on relativity in 1908. Minkowsky was the maths teacher to the more-famous Einstein. Minkowsky realized that inputs from the human senses, which seem subjectively to represent a spatial three-dimensional world, are mapped on to the inseparable entity of the moments of time to form a higher four-dimensional reality (spacetime). Einstein, interestingly, did not agree with Minkowsky, but later incorporated Minkowsky's ideas into his own theories of how time is woven into space. Einstein was the rock star of the timespace relativity principle, and he makes a great quote, but his work was grounded, as are all advances, on the work of other visionary physicists.

8. Stress

i. CRH is carried in small blood vessels from the hypothalamus, near the frontal part of the base of the brain, to the pituitary gland that sits in a bony indentation just below it but outside the brain. Here, CRH stimulates the release of ACTH. ACTH is carried in the blood circulation and brings about the release of cortisol from the adrenal glands that lie on top of the kidneys. Cortisol is released into the blood stream and brought around to different organs in the body, where it causes changes in DNA production intracellularly (in individual cells) and alters the protein production in these cells.

ii. Bruce McEwan died in January 2020. In 2019 I had the greatest of pleasures, after a decade of admiration, of his co-authoring an editorial in *Biological Psychiatry* on my MRI paper on hippocampal size in depression.[20] He worked right up until his death aged eighty-two years.

9. Self-Recognition

i. Everything about us exemplifies how we humans have embodied our phylogenetic ancestors. The evidence of this seeps out in a particularly spectacular way during fetal development. Although it is too crude to say that the human embryo as it develops mirrors our phylogenetic development from our ancestors, there are key points in embryonic development where a structure could theoretically develop into several different species. The structure could, for example, evolve into the gills of a fish, a huge reptilian jaw or the ear, nose and throat of a human. This area of science is known as evo-devo (evolutionary developmental biology), and it examines the development of life from embryo to adult as it has evolved genetically.

ii. One of the most amazing things about the brain is the existence of big neurons that are only present in the anterior insula, the cingula, and the extreme prefrontal cortex, on the frontal edge of the brain. The neurons in this circuit were first described, before there was any inkling of their functional importance, by a psychiatrist and neurologist called Constantin von Economo back in 1929. The neurons had a specific and quite primitive design – big, long and with simple connections – and are often now referred to as 'spindle' neurons. These von Economo neurons (VENs) allow emotional signals from the body mapped on to the insula to be projected in real-life time, like an emotional projector, to the integrative prefrontal brain. The VENs give the prefrontal integrative brain moment-to-moment awareness of an emotional state.

A hint of the importance of the VENs in the experience of being human is that they are only present in a limited number of younger species – humans, some apes, elephants, some whales and dolphins – all of whom have advanced forms of self-awareness. Animals who possess VENs pass the mirror self-awareness test. Consistent with the pattern of infant human development mirroring that of evolutionary development, VENs only appear in the later stages of gestation and grow during infancy – the period when infants develop self-awareness.

10. The Tree of Life

i. There is some uncertainty about whether Beethoven was fully without hearing during his thirties and forties, and this was a very sensitive issue for him. He denied his hearing impairment, but his deafness was evident from his problems with hearing others and with his difficulty hearing live music.

ii. Myelin is formed from non-neuronal cells, adjacent to neurons, that also form a scaffold for the neurons.

iii. An example of autoimmune disease is NMDA encephalitis, as discussed in Chapter 2 (p. 17). Inflammatory markers, such as cytokines, are probably more prevalent in depression.

11. A Sense of Self

i. I have taken the phrase 'narrative consciousness' from *Amnesiac Selves: Nostalgia, Forgetting, and British Fiction, 1810–1870* (Oxford University Press, 2001), by Nicholas Dames. He is Theodore Kahan Professor of Humanities in Columbia University and has written extensively on the development of modern psychology from the Victorian novel in the nineteenth century to modern literature.

ii. Psychedelics are now being tested for possible therapeutic use in a variety of psychiatric disorders. There is good evidence for the use of ketamine in depression, although it remains to be seen whether the anti-depressant effects will be long-lasting in individual patients. I am involved in the first international multi-centre trials of using psilocybin – the psychoactive compound in magic mushrooms – in treating severe depression.[157]

12. Sex Hormones and Songbirds

i. One theory, which incidentally was the subject matter of my PhD in the early 1990s, was that estrogen and progesterone may affect feeling circuitry by bringing about changes in brain neurotransmitter function.[158, 159] We were interested in the possible effects of estrogen and progesterone on serotonin function, because the drugs that are used to treat depression increase serotonin function. My work is now dated, but the theory has been investigated with more sophisticated methods in subsequent years. A group from Copenhagen has looked at brain serotonin neuroimaging function in women, measured through labelling serotonin with radioactive atoms and then looking at the patterns of illuminated serotonin.[160] They reported that fluctuations in estrogen levels cause changes in the seretonin-mediated circuitry, in both the amygdala and the hippocampus.

ii. The effect of estrogens on the emotional circuitry probably underlies the heightened emotionality commonly experienced during periods of fluctuating estrogen hormone levels – adolescence, pregnancy and following childbirth, and during the menopause.[74] This effect is reflected in the mood disorders in women that cluster around times of fluctuating female sex hormones.

iii. A lobotomy, also known as a leucotomy, is a neurosurgical procedure where connections to the prefrontal cortex are severed. It was quite a common procedure to treat serious mental illness in the 1950s, before the discovery of psychiatric drugs. Such neurosurgery for the treatment of psychiatric disorders is very rarely used.

14. *False or True?*

i. Bentley's paper from 1899 can be read in full at https://www.jstor.org/stable/pdf/1412727. Many of the concepts theorized about in his paper could be translated into a psychology journal today. The language has changed, but there is a freshness of thought and an expansiveness in manuscripts from this period because of the freedom from today's specialization within the neurosciences. The physicist James Clerk Maxwell put this very well twenty years prior to the Bentley paper, when he wrote, 'It is of great advantage to the student of any subject to read the original memoirs on that subject, for science is always most completely assimilated when it is in the nascent state.' (From the Preface to *A Treatise on Electricity and Magnetism* (1873).)

ii. The following quote describes Freud's conversion from the idea that sexual abuse was a recalled memory to the idea that the memory was an unconscious fantasy based on infantile sexuality, that is, attraction of the girl child to the father:

> the neurotic symptoms were not related directly to actual events but to wishful fantasies, and that as far as the neurosis was concerned psychical reality was of more importance than material reality. I do not believe even now that I forced the seduction-fantasies on my patients, or that I 'suggested' them. I had in fact stumbled for the first time upon the Oedipus complex, which was later to assume such an overwhelming importance, but which I did not recognize as yet in its disguise of fantasy . . . When the mistake had been cleared up, the path to the study of the sexual life of children lay open. (*An Autobiographical Study* (1925): http://www.mhweb.org/mpc_course/freud.pdf)

Notes

iii. Tonegawa worked out one of the most difficult puzzles in immunology – how an immune cell makes multiple antibodies – and won the Nobel Prize in Physiology in 1987 for this discovery. A few years later, in 1990, he changed tack and steered his restless razor-sharp mind to one of the most challenging questions in neuroscience: the puzzle of memory. How a limited number of hippocampal cells can make, store and recall so many memories is perhaps not such a different puzzle from that of how immune cells produce such a diverse range of antibodies. You may remember that in the 1970s and 1980s Edelman made the same transition from immunity to neuroscience, perhaps reflecting the fact that immunity requires cellular memory.

iv. Channel rhodopsins (what I refer to in the text as rhodopsins) were first isolated by Georg Nagel's lab in the Max Planck Institute, where scientists were searching for the proteins that cause photocurrents in the unicellular green alga. Then Nagel collaborated with Edward Boyden (at the time a PhD student at Stanford University, California) and Karl Deisseroth (also from Stanford University) and used this method to tag on to the dormant rhodopsin-like DNA in neurons and produce channel rhodopsins. Remarkably, this feat of genetic engineering was predicted by Francis Crick – the Crick of Crick and Watson fame, who unwound the hidden helical structure of DNA – in a series of lectures in 1999 at the University of California in San Diego.

v. Tomás Ryan also thinks that this was not a false, but a manipulated, memory. The term was being used loosely within the generally confused understanding of what constitutes a false memory.

15. The Oldest Memories

i. The human genome project laid out all of the genes in the human cell, 30,000 or so, which are collectively called the genome. It was the biggest scientific collaboration up to that time, commencing in 1990 and completed ahead of its anticipated finishing date in 2003. All living organisms are derived from cells that store genetic information using the same huge DNA molecule. DNA is stored in a complex ladder-like zip structure, twisted on its own axis, and then scrunched up into a bundle. It is all composed of four molecules that contain the code for life: A, C, G and T. There are three billion of these 'letters' in the human genome, organized in sequences that form unique codes for each protein, and most of these proteins are common to *all* forms of life. Approximately 99 per cent of our working DNA is the same as that of chimpanzees. We share 60 per cent of our DNA with that of the banana.

One of the most striking scientific stories recently is about the micro-organisms that inhabit one's gut, called the gut microbiome. The story of how the microbiome in the human gut can influence brain function involves my good friend Ted Dinan, a psychiatrist and brain researcher from University College Cork, with whom I did my PhD, and John Cryan, a lively and clever scientist. The gut microbiome consists of about 100 trillion cells – about three times more cells than the human body – that work in a symbiotic way with the body. Humans have incorporated, literally, these simple microorganisms unchanged, and they have been transformed into a living human physiological system that lives side-by-side with other human physiological systems.[161]

ii. From 'Plasticity and the Ageing Mind: An Exemplar of the Bio-cultural Orchestration of Brain and Behaviour', *European Review*, 9:1 (2001), 59–76.

iii. I fell, delighted, upon the re-publication in 2014 of the original version of the *Brothers Grimm Folk and Fairy Tales* (Princeton University Press, 2014). These pithy stories are shockingly different from the ones that children read and are certainly only for adult consumption. They demonstrate that these stories reflect real life, rather than nice fantasies. Charles Perrault's collection of fairy tales from France that he put together at the end of the seventeenth century is the first European collection of oral folk tales and has been translated in its authentic rawness. (Oxford, 2009). One notorious story is about a man who is a serial wife murderer (Bluebeard). Many stories involve incest and infanticide.

iv. One of my favourite books when I was a child was a collection of fairy stories by Sinead de Valera (*Irish Fairy Stories* [Pan Books, 1973]), who was married to the third president of Ireland, Eamon de Valera, also a revolutionary in the War of Independence.

v. The delusion of substitution has a special name, Capgras syndrome, after the French psychiatrist (1873–1950) who first identified it. It is more frequently associated with cognitive impairment in the elderly, and not with postpartum psychosis.

vi. This is a quotation from Erin Kraus's book *Wise-woman of Kildare: Moll Anthony and Popular Tradition in the East of Ireland* (Four Courts, 2011). The book is about 'wise women', or fairy doctors, who were renowned locally in Ireland.

References

1. Kuppuswamy P. S., Takala C. R., Sola C. L., 'Management of psychiatric symptoms in anti-NMDAR encephalitis: a case series, literature review and future directions', *General Hospital Psychiatry*, 2014; 36:388–91.
2. Sansing L. H., Tuzun E., Ko M. W., Baccon J., Lynch D. R., Dalmau J., 'A patient with encephalitis associated with NMDA receptor antibodies', *Nature Clinical Practice Neurology*, 2007; 3:291–6.
3. Jezequel J., Johansson E. M., Dupuis J. P., et al., 'Dynamic disorganization of synaptic NMDA receptors triggered by autoantibodies from psychotic patients', *Nature Communications*, 2017; 8:1791.
4. Bassett D. S., Sporns O., 'Network neuroscience', *Nature Neuroscience*, 2017; 20:353–64.
5. Scott J., Martin G., Bor W., Sawyer M., Clark J., McGrath J., 'The prevalence and correlates of hallucinations in Australian adolescents: results from a national survey', *Schizophrenia Research*, 2009; 107:179–85.
6. van Os J., Linscott R. J., Myin-Germeys I., Delespaul P., Krabbendam L., 'A systematic review and meta-analysis of the psychosis continuum: evidence for a psychosis proneness–persistence–impairment model of psychotic disorder', *Psychological Medicine*, 2009; 39:179–95.
7. Kurth R., Villringer K., Curio G., et al., 'fMRI shows multiple somatotopic digit representations in human primary somatosensory cortex', *Neuro-Report*, 2000; 11:1487–91.
8. Ortiz-Teran L., Ortiz T., Perez D. L., et al., 'Brain plasticity in blind subjects centralizes beyond the modal cortices', *Frontiers in Systems Neuroscience*, 2016; 10:61.
9. Haigh A., Brown D. J., Meijer P., Proulx M. J., 'How well do you see what you hear? The acuity of visual-to-auditory sensory substitution', *Frontiers in Psychology*, 2013; 4:330.
10. Berger J., 'Raising the portcullis: some notes after having cataracts removed from my eyes', *British Journal of General Practice*, 2010; 60:464–5.
11. Shergill S. S., Brammer M. J., Williams S. C., Murray R. M., McGuire P. K., 'Mapping auditory hallucinations in schizophrenia using functional magnetic resonance imaging', *Archive of General Psychiatry*, 2000; 57:1033–8.

References

12. Plaze M., Paillere-Martinot M. L., Penttila J., et al., ' "Where do auditory hallucinations come from?" – a brain morphometry study of schizophrenia patients with inner or outer space hallucinations', *Schizophrenia Bulletin*, 2011; 37:212–21.

13. Luo Y., He H., Duan M., et al., 'Dynamic functional connectivity strength within different frequency-band in schizophrenia', *Frontiers in Psychiatry*, 2019; 10:995.

14. Hurst L. C., 'What was wrong with Anna O?', *Journal of the Royal Society of Medicine*, 1982; 75:129–31.

15. Slater E. T., Glithero E., 'A follow-up of patients diagnosed as suffering from "hysteria" ', *Journal of Psychosomatic Research*, 1965; 9:9–13.

16. Scoville W. B., Milner B., 'Loss of recent memory after bilateral hippocampal lesions', *Journal of Neurology, Neurosurgery and Psychiatry*, 1957; 20:11–21.

17. Vargha-Khadem F., Gadian D. G., Watkins K. E., Connelly A., Van Paesschen W., Mishkin M., 'Differential effects of early hippocampal pathology on episodic and semantic memory', *Science* 1997; 277:376–80.

18. Maguire E. A., Gadian D. G., Johnsrude I. S., et al., 'Navigation-related structural change in the hippocampi of taxi drivers', *Proceedings of the National Academy of Sciences of the USA*, 2000; 97:4398–403.

19. Schmaal L., Veltman D. J., van Erp T. G., et al., 'Subcortical brain alterations in major depressive disorder: findings from the ENIGMA Major Depressive Disorder working group', *Molecular Psychiatry*, 2016; 21:806–12.

20. Roddy D. W., Farrell C., Doolin K., et al., 'The hippocampus in depression: more than the sum of its parts? Advanced hippocampal substructure segmentation in depression', *Biological Psychiatry*, 2019; 85:487–97.

21. Viard A., Piolino P., Desgranges B., et al., 'Hippocampal activation for autobiographical memories over the entire lifetime in healthy aged subjects: an fMRI study', *Cerebral Cortex*, 2007; 17:2453–67.

22. Squire L. R., Alvarez P., 'Retrograde amnesia and memory consolidation: a neurobiological perspective', *Current Opinion in Neurobiology*, 1995; 5:169–77.

23. Daselaar S. M., Rice H. J., Greenberg D. L., Cabeza R., LaBar K. S., Rubin D. C., 'The spatiotemporal dynamics of autobiographical memory: neural correlates of recall, emotional intensity, and reliving', *Cerebral Cortex*, 2008; 18:217–29.

24. Preston A. R., Eichenbaum H., 'Interplay of hippocampus and prefrontal cortex in memory', *Current Biology*, 2013; 23:R764–73.

25. Piefke M., Weiss P. H., Zilles K., Markowitsch H. J., Fink G. R., 'Differential remoteness and emotional tone modulate the neural correlates of autobiographical memory', *Brain*, 2003; 126:650–68.

26. Wamsley E. J., 'Rhythms of sleep: orchestrating memory consolidation (commentary on Clemens et al.)', *European Journal of Neuroscience*, 2011; 33:509–10.

27. Batterink L. J., Creery J. D., Paller K. A., 'Phase of spontaneous slow oscillations during sleep influences memory-related processing of auditory cues', *Journal of Neuroscience*, 2016; 36:1401–9.

28. de Sousa A. F., Cowansage K. K., Zutshi I., et al., 'Optogenetic reactivation of memory ensembles in the retrosplenial cortex induces systems consolidation', *Proceedings of the National Academy of Sciences of the USA*, 2019; 116:8576–81.

29. Buck L. B., 'Olfactory receptors and odor coding in mammals', *Nutritional Reviews*, 2004; 62:S184–8; discussion S224–41.

30. Siebert M., Markowitsch H. J., Bartel P., 'Amygdala, affect and cognition: evidence from 10 patients with Urbach–Wiethe disease', *Brain*, 2003; 126:2627–37.

31. Bechara A., Tranel D., Damasio H., Adolphs R., Rockland C., Damasio A. R., 'Double dissociation of conditioning and declarative knowledge relative to the amygdala and hippocampus in humans', *Science*, 1995; 269:1115–18.

32. Adolphs R., Tranel D., Damasio H., Damasio A., 'Impaired recognition of emotion in facial expressions following bilateral damage to the human amygdala', *Nature*, 1994; 372:669–72.

33. Feinstein J. S., Adolphs R., Damasio A., Tranel D., 'The human amygdala and the induction and experience of fear', *Current Biology*, 2011; 21:34–8.

34. Phelps E. A., LeDoux J. E., 'Contributions of the amygdala to emotion processing: from animal models to human behavior', *Neuron*, 2005; 48:175–87.

35. Dilger S., Straube T., Mentzel H. J., et al., 'Brain activation to phobia-related pictures in spider phobic humans: an event-related functional magnetic resonance imaging study', *Neuroscience Letters*, 2003; 348:29–32.

36. James W., 'The physical basis of emotion', *Psychological Review*, 1894; 101:205–10.

37. Craig A. D., 'How do you feel – now? The anterior insula and human awareness', *Nature Reviews Neuroscience*, 2009; 10:59–70.

38. Verstaen A., Eckart J. A., Muhtadie L., et al., 'Insular atrophy and diminished disgust reactivity', *Emotion*, 2016; 16:903–12.

39. Ehrlich S., Lord A. R., Geisler D., et al., 'Reduced functional connectivity in the thalamo-insular subnetwork in patients with acute anorexia nervosa', *Human Brain Mapping*, 2015; 36:1772–81.

40. Surguladze S. A., El-Hage W., Dalgleish T., Radua J., Gohier B., Phillips M. L., 'Depression is associated with increased sensitivity to signals of disgust: a functional magnetic resonance imaging study', *Journal of Psychiatric Research*, 2010; 44:894–902.

41. Penfield W., Faulk M. E., Jr, 'The insula: further observations on its function', *Brain*, 1955; 78:445–70.

42. Nguyen D. K., Nguyen D. B., Malak R., et al., 'Revisiting the role of the insula in refractory partial epilepsy', *Epilepsia*, 2009; 50:510–20.

References

43. Critchley H. D., Wiens S., Rotshtein P., Ohman A., Dolan R. J., 'Neural systems supporting interoceptive awareness', *Nature Neuroscience*, 2004; 7:189–95.
44. Critchley H. D., Tang J., Glaser D., Butterworth B., Dolan R. J., 'Anterior cingulate activity during error and autonomic response', *NeuroImage*, 2005; 27:885–95.
45. Knutson B., Rick S., Wimmer G. E., Prelec D., Loewenstein G., 'Neural predictors of purchases', *Neuron*, 2007; 53:147–56.
46. Namkung H., Kim S. H., Sawa A., 'The insula: an underestimated brain area in clinical neuroscience, psychiatry, and neurology', *Trends in Neurosciences*, 2017; 40:200–207.
47. Eisenberger N. I., Lieberman M. D., Williams K. D., 'Does rejection hurt? An FMRI study of social exclusion', *Science*, 2003; 302:290–92.
48. O'Keefe J., Dostrovsky J., 'The hippocampus as a spatial map. Preliminary evidence from unit activity in the freely-moving rat', *Brain Research*, 1971; 34: 171–5.
49. Colgin L. L., Moser E. I., Moser M. B., 'Understanding memory through hippocampal remapping', *Trends in Neurosciences*, 2008; 31:469–77.
50. Ekstrom A. D., Kahana M. J., Caplan J. B., et al., 'Cellular networks underlying human spatial navigation', *Nature*, 2003; 425:184–8.
51. Maguire E. A., Mummery C. J., 'Differential modulation of a common memory retrieval network revealed by positron emission tomography', *Hippocampus*, 1999; 9:54–61.
52. Rowland D. C., Roudi Y., Moser M. B., Moser E. I., 'Ten years of grid cells', *Annual Review of Neuroscience*, 2016; 39:19–40.
53. Hafting T., Fyhn M., Bonnevie T., Moser M. B., Moser E. I., 'Hippocampus-independent phase precession in entorhinal grid cells', *Nature*, 2008; 453: 1248–52.
54. Jacobs J., Weidemann C. T., Miller J. F., et al., 'Direct recordings of grid-like neuronal activity in human spatial navigation', *Nature Neuroscience*, 2013; 16: 1188–90.
55. Hall J., Thomas K. L., Everitt B. J., 'Cellular imaging of zif268 expression in the hippocampus and amygdala during contextual and cued fear memory retrieval: selective activation of hippocampal CA1 neurons during the recall of contextual memories', *Journal of Neuroscience*, 2001; 21:2186–93.
56. Horowitz J. M., Horwitz B. A., 'Extreme neuroplasticity of hippocampal CA1 pyramidal neurons in hibernating mammalian species', *Frontiers in Neuroanatomy*, 2019; 13:9.
57. Eichenbaum H., 'Memory on time', *Trends in Cognitive Sciences*, 2013; 17:81–8.
58. Tsao A., Sugar J., Lu L., et al., 'Integrating time from experience in the lateral entorhinal cortex', *Nature*, 2018; 561:57–62.

59. MacDonald C. J., Lepage K. Q., Eden U. T., Eichenbaum H., 'Hippocampal "time cells" bridge the gap in memory for discontiguous events', *Neuron*, 2011; 71:737–49.

60. Deuker L., Bellmund J. L., Navarro Schroder T., Doeller C. F., 'An event map of memory space in the hippocampus', *eLife*, 2016; 5.

61. Manning L., Cassel D., Cassel J. C., 'St. Augustine's reflections on memory and time and the current concept of subjective time in mental time travel', *Behavioral Sciences* (Basel), 2013; 3:232–43.

62. Rosenbaum R. S., Kohler S., Schacter D. L., et al., 'The case of K.C.: contributions of a memory-impaired person to memory theory', *Neuropsychologia*, 2005; 43:989–1021.

63. Addis D. R., Pan L., Vu M. A., Laiser N., Schacter D. L., 'Constructive episodic simulation of the future and the past: distinct subsystems of a core brain network mediate imagining and remembering', *Neuropsychologia*, 2009; 47:2222–38.

64. Buckner R. L., Andrews-Hanna J. R., Schacter D. L., 'The brain's default network: anatomy, function, and relevance to disease', *Annals of the NY Academy of Sciences*, 2008; 1124:1–38.

65. Addis D. R., Sacchetti D. C., Ally B. A., Budson A. E., Schacter D. L., 'Episodic simulation of future events is impaired in mild Alzheimer's disease', *Neuropsychologia*, 2009; 47:2660–71.

66. Moskowitz A. K., ' "Scared stiff": catatonia as an evolutionary-based fear response', *Psychological Review*, 2004; 111:984–1002.

67. Lupien S. J., Wilkinson C. W., Briere S., Menard C., Ng Ying Kin N. M., Nair N. P., 'The modulatory effects of corticosteroids on cognition: studies in young human populations', *Psychoneuroendocrinology*, 2002; 27:401–16.

68. Juster R. P., McEwen B. S., Lupien S. J., 'Allostatic load biomarkers of chronic stress and impact on health and cognition', *Neuroscience and Biobehavioral Reviews*, 2010; 35:2–16.

69. Pariante C. M., Lightman S. L., 'The HPA axis in major depression: classical theories and new developments', *Trends in Neurosciences*, 2008; 31:464–8.

70. Cleare A. J., Bearn J., Allain T., et al., 'Contrasting neuroendocrine responses in depression and chronic fatigue syndrome', *Journal of Affective Disorders*, 1995; 34:283–9.

71. Sarrieau A., Vial M., McEwen B., et al., 'Corticosteroid receptors in rat hippocampal sections: effect of adrenalectomy and corticosterone replacement', *Journal of Steroid Biochemistry*, 1986; 24:721–4.

72. de Kloet E. R., Joels M., Holsboer F., 'Stress and the brain: from adaptation to disease', *Nature Reviews Neuroscience*, 2005; 6:463–75.

References

73. Joels M., de Kloet E. R., 'Effects of glucocorticoids and norepinephrine on the excitability in the hippocampus', *Science*, 1989; 245:1502–5.

74. O'Keane V., Lightman S., Patrick K., Marsh M., Papadopoulos A. S., Pawlby S., Seneviratne G., Taylor A., Moore R. J., 'Changes in the maternal hypothalamic–pituitary–adrenal axis during the early puerperium may be related to the postpartum "blues"', *Neuroendocrinology*, 2011; 11:1149–55.

75. Meaney M. J., Aitken D. H., Bodnoff S. R., Iny L. J., Sapolsky R. M., 'The effects of postnatal handling on the development of the glucocorticoid receptor systems and stress recovery in the rat', *Progress in Neuropsychopharmacology and Biological Psychiatry*, 1985; 9:731–4.

76. Magarinos A. M., Verdugo J. M., McEwen B. S., 'Chronic stress alters synaptic terminal structure in hippocampus', *Proceedings of the National Academy of Sciences of the USA*, 1997; 94:14002–8.

77. McEwen B. S., 'Allostasis and allostatic load: implications for neuropsychopharmacology', *Neuropsychopharmacology*, 2000; 22:108–24.

78. Ouellet-Morin I., Robitaille M. P., Langevin S., Cantave C., Brendgen M., Lupien S. J., 'Enduring effect of childhood maltreatment on cortisol and heart rate responses to stress: the moderating role of severity of experiences', *Development and Psychopathology*, 2019; 31:497–508.

79. Frodl T., O'Keane V., 'How does the brain deal with cumulative stress? A review with focus on developmental stress, HPA axis function and hippocampal structure in humans', *Neurobiology of Disease*, 2013; 52:24–37.

80. Warner-Schmidt J. L., Duman R. S., 'Hippocampal neurogenesis: opposing effects of stress and antidepressant treatment', *Hippocampus*, 2006; 16: 239–49.

81. Tozzi L., Doolin K., Farrel C., Joseph S., O'Keane V., Frodl T., 'Functional magnetic resonance imaging correlates of emotion recognition and voluntary attentional regulation in depression: A generalized psycho-physiological interaction study', *Journal of Affective Disorders*, 2017; 208:535–44.

82. Frodl T., Strehl K., Carballedo A., Tozzi L., Doyle M., Amico F., Gormley J., Lavelle G., O'Keane V., 'Aerobic exercise increases hippocampal subfield volumes in younger adults and prevents volume decline in the elderly', *Brain Imaging and Behaviour*, March 2019.

83. Tozzi L., Carballedo A., Lavelle G., Doolin K., Doyle M., Amico F., McCarthy H., Gormley J., Lord A., O'Keane V., Frodl T., 'Longitudinal functional connectivity changes correlate with mood improvement after regular exercise in a dose-dependent fashion', *European Journal of Neuroscience* 2016; 43(8): 1089–96.

84. Carroll S. B., 'Evo-devo and an expanding evolutionary synthesis: a genetic theory of morphological evolution', *Cell*, 2008; 134:25–36.

85. Lewis M., Ramsay D., 'Development of self-recognition, personal pronoun use, and pretend play during the 2nd year', *Child Development*, 2004; 75: 1821–31.

86. Plotnik J. M., de Waal F. B., Reiss D., 'Self-recognition in an Asian elephant', *Proceedings of the National Academy of Sciences of the USA*, 2006; 103:17053–7.

87. Prior H., Schwarz A., Gunturkun O., 'Mirror-induced behavior in the magpie (Pica pica): evidence of self-recognition', *PLoS Biology*, 2008; 6:e202.

88. Hutchison W. D., Davis K. D., Lozano A. M., Tasker R. R., Dostrovsky J. O., 'Pain-related neurons in the human cingulate cortex', *Nature Neuroscience*, 1999; 2:403–5.

89. Swiney L., Sousa P., 'A new comparator account of auditory verbal hallucinations: how motor prediction can plausibly contribute to the sense of agency for inner speech', *Frontiers in Human Neuroscience*, 2014; 8:675.

90. Bastiaansen J. A., Thioux M., Keysers C., 'Evidence for mirror systems in emotions', *Philosophical Transactions of the Royal Society of London Series B: Biological Sciences*, 2009; 364:2391–404.

91. Carr L., Iacoboni M., Dubeau M. C., Mazziotta J. C., Lenzi G. L., 'Neural mechanisms of empathy in humans: a relay from neural systems for imitation to limbic areas', *Proceedings of the National Academy of Sciences of the USA*, 2003; 100:5497–502.

92. Singer T., Seymour B., O'Doherty J., Kaube H., Dolan R. J., Frith C. D., 'Empathy for pain involves the affective but not sensory components of pain', *Science*, 2004; 303:1157–62.

93. Meffert H., Gazzola V., den Boer J. A., Bartels A. A., Keysers C., 'Reduced spontaneous but relatively normal deliberate vicarious representations in psychopathy', *Brain*, 2013; 136:2550–62.

94. Wiech K., Jbabdi S., Lin C. S., Andersson J., Tracey I., 'Differential structural and resting state connectivity between insular subdivisions and other pain-related brain regions', *Pain*, 2014; 155:2047–55.

95. Butti C., Hof P. R., 'The insular cortex: a comparative perspective', *Brain Structure and Function*, 2010; 214:477–93.

96. Seeley W. W., Carlin D. A., Allman J. M., et al., 'Early frontotemporal dementia targets neurons unique to apes and humans', *Annals of Neurology*, 2006; 60:660–67.

97. Allman J. M., Watson K. K., Tetreault N. A., Hakeem A. Y., 'Intuition and autism: a possible role for Von Economo neurons', *Trends in Cognitive Sciences*, 2005; 9:367–73.

98. Dolan R. J., Fletcher P. C., McKenna P., Friston K. J., Frith C. D., 'Abnormal neural integration related to cognition in schizophrenia', *Acta Psychiatrica Scandinavica*, 1999; s395:58–67.

99. Brune M., Schobel A., Karau R., et al., 'Von Economo neuron density in the anterior cingulate cortex is reduced in early onset schizophrenia', *Acta Neuropathologica*, 2010; 119:771–8.

100. Costain G., Ho A., Crawley A. P., et al., 'Reduced gray matter in the anterior cingulate gyrus in familial schizophrenia: a preliminary report', *Schizophrenia Research*, 2010; 122:81–4.

101. Rizzolatti G., 'Multiple body representations in the motor cortex of primates', *Acta Biomedica Ateneo Parmense*, 1992; 63:27–9.

102. Maranesi M., Livi A., Fogassi L., Rizzolatti G., Bonini L., 'Mirror neuron activation prior to action observation in a predictable context', *Journal of Neuroscience*, 2014; 34:14827–32.

103. Herholz S. C., Halpern A. R., Zatorre R. J., 'Neuronal correlates of perception, imagery, and memory for familiar tunes', *Journal of Cognitive Neuroscience*, 2012; 24:1382–97.

104. Gogtay N., Giedd J. N., Lusk L., et al., 'Dynamic mapping of human cortical development during childhood through early adulthood', *Proceedings of the National Academy of Sciences of the USA*, 2004; 101:8174–9.

105. Zhou D., Lebel C., Treit S., Evans A., Beaulieu C., 'Accelerated longitudinal cortical thinning in adolescence', *NeuroImage*, 2015; 104:138–45.

106. Storsve A. B., Fjell A. M., Tamnes C. K., et al., 'Differential longitudinal changes in cortical thickness, surface area and volume across the adult life span: regions of accelerating and decelerating change', *Journal of Neuroscience*, 2014; 34:8488–98.

107. Boksa P., 'Abnormal synaptic pruning in schizophrenia: Urban myth or reality?', *Journal of Psychiatry and Neuroscience*, 2012; 37:75–7.

108. Whitaker K. J., Vertes P. E., Romero-Garcia R., et al., 'Adolescence is associated with genomically patterned consolidation of the hubs of the human brain connectome', *Proceedings of the National Academy of Sciences of the USA*, 2016; 113:9105–10.

109. O'Callaghan E., Sham P., Takei N., Glover G., Murray R. M., 'Schizophrenia after prenatal exposure to 1957 A2 influenza epidemic', *Lancet*, 1991; 337:1248–50.

110. Murray R. M., 'Mistakes I have made in my research career', *Schizophrenia Bulletin*, 2017; 43:253–6.

111. Weinstock M., 'Alterations induced by gestational stress in brain morphology and behaviour of the offspring', *Progress in Neurobiology*, 2001; 65:427–51.

112. Salat D. H., Buckner R. L., Snyder A. Z., et al., 'Thinning of the cerebral cortex in aging', *Cerebral Cortex*, 2004; 14:721–30.

113. Elliott B., Joyce E., Shorvon S., 'Delusions, illusions and hallucinations in epilepsy: 2. Complex phenomena and psychosis', *Epilepsy Research*, 2009; 85:172–86.

114. Edelman G. M., Gally J. A., 'A model for the 7s antibody molecule', *Proceedings of the National Academy of Sciences of the USA*, 1964; 51:846–53.

115. Hall Z. J., Macdougall-Shackleton S. A., 'Influence of testosterone metabolites on song-control system neuroplasticity during photostimulation in adult European starlings (*Sturnus vulgaris*)', *PLoS One*, 2012; 7:e40060.

116. Draper P., Belsky J., 'Personality development in the evolutionary perspective', *Journal of Personality*, 1990; 58:141–61.

117. Plant T. M., 'The role of KiSS-1 in the regulation of puberty in higher primates', *European Journal of Endocrinology*, 2006; 155 Suppl 1:S11–16.

118. Ball G. F., Ketterson E. D., 'Sex differences in the response to environmental cues regulating seasonal reproduction in birds', *Philosophical Transactions of the Royal Society of London Series B: Biological Sciences*, 2008; 363:231–46.

119. Bean L. A., Ianov L., Foster T. C., 'Estrogen receptors, the hippocampus, and memory', *Neuroscientist*, 2014; 20:534–45.

120. Wierckx K., Elaut E., Van Hoorde B., et al., 'Sexual desire in trans persons: associations with sex reassignment treatment', *Journal of Sexual Medicine*, 2014; 11:107–18.

121. Hamann S., Stevens J., Vick J. H., et al., 'Brain responses to sexual images in 46,XY women with complete androgen insensitivity syndrome are female-typical', *Hormones and Behavior*, 2014; 66:724–30.

122. Henningsson S., Madsen K. H., Pinborg A., et al., 'Role of emotional processing in depressive responses to sex-hormone manipulation: a pharmacological fMRI study', *Translational psychiatry*, 2015; 5:e688.

123. Miedl S. F., Wegerer M., Kerschbaum H., Blechert J., Wilhelm F. H., 'Neural activity during traumatic film viewing is linked to endogenous estradiol and hormonal contraception', *Psychoneuroendocrinology*, 2018; 87:20–26.

124. Sotres-Bayon F., Bush D. E., LeDoux J. E., 'Emotional perseveration: an update on prefrontal-amygdala interactions in fear extinction', *Learning and Memory*, 2004; 11:525–35.

125. Choudhury S., Blakemore S. J., Charman T., 'Social cognitive development during adolescence', *Social Cognitive and Affective Neuroscience*, 2006; 1:165–74.

126. Teicher M. H., Samson J. A., 'Annual research review: enduring neurobiological effects of childhood abuse and neglect', *Journal of Child Psychology and Psychiatry*, 2016; 57:241–66.

127. De Bellis M. D., Keshavan M. S., Shifflett H., et al., 'Brain structures in pediatric maltreatment-related posttraumatic stress disorder: a socio-demographically matched study', *Biological Psychiatry*, 2002; 52:1066–78.

128. Pechtel P., Lyons-Ruth K., Anderson C. M., Teicher M. H., 'Sensitive periods of amygdala development: the role of maltreatment in preadolescence', *NeuroImage*, 2014; 97:236–44.

129. Whittle S., Dennison M., Vijayakumar N., et al., 'Childhood maltreatment and psychopathology affect brain development during adolescence', *Journal of the American Academy of Child and Adolescent Psychiatry*, 2013; 52:940–52 e1.

130. Cullen K. R., Vizueta N., Thomas K. M., et al., 'Amygdala functional connectivity in young women with borderline personality disorder', *Brain Connectivity*, 2011; 1:61–71.

131. Linehan M. M., Heard H. L., Armstrong H. E., 'Naturalistic follow-up of a behavioral treatment for chronically parasuicidal borderline patients', *Archive of General Psychiatry*, 1993; 50:971–4.

132. Stoffers J. M., Vollm B. A., Rucker G., Timmer A., Huband N., Lieb K., 'Psychological therapies for people with borderline personality disorder', *Cochrane Database of Systematic Reviews*, 2012:CD005652.

133. Roberts B. W., Caspi A., Moffitt T. E., 'The kids are alright: growth and stability in personality development from adolescence to adulthood', *Journal of Personality and Social Psychology*, 2001; 81:670–83.

134. Goodman M., Carpenter D., Tang C. Y., et al., 'Dialectical behavior therapy alters emotion regulation and amygdala activity in patients with borderline personality disorder', *Journal of Psychiatric Research*, 2014; 57:108–16.

135. Mundt A. P., Chow W. S., Arduino M., et al., 'Psychiatric hospital beds and prison populations in South America since 1990: does the Penrose hypothesis apply?', *JAMA Psychiatry*, 2015; 72:112–18.

136. Gulati G., Keating N., O'Neill A., Delaunois I., Meagher D., Dunne C. P., 'The prevalence of major mental illness, substance misuse and homelessness in Irish prisoners: systematic review and meta-analyses', *Irish Journal of Psychological Medicine*, 2019; 36:35–45.

137. Follette V. M., Polusny M. A., Bechtle A. E., Naugle A. E., 'Cumulative trauma: the impact of child sexual abuse, adult sexual assault, and spouse abuse', *Journal of Traumatic Stress*, 1996; 9:25–35.

138. Cochran K. J., Greenspan R. L., Bogart D. F., Loftus E. F., 'Memory blindness: altered memory reports lead to distortion in eyewitness memory', *Memory and Cognition*, 2016; 44:717–26.

139. Deisseroth K., 'Optogenetics', *Nature Methods*, 2011; 8:26–9.

140. Bi A., Cui J., Ma Y. P., et al., 'Ectopic expression of a microbial-type rhodopsin restores visual responses in mice with photoreceptor degeneration', *Neuron*, 2006; 50:23–33.

141. Nagel G., Ollig D., Fuhrmann M., et al., 'Channelrhodopsin-1: a light-gated proton channel in green algae', *Science*, 2002; 296:2395–8.

142. Boyden E. S., Zhang F., Bamberg E., Nagel G., Deisseroth K., 'Millisecond-timescale, genetically targeted optical control of neural activity', *Nature Neuroscience*, 2005; 8:1263–8.

143. Ramirez S., Liu X., Lin P. A., et al., 'Creating a false memory in the hippocampus', *Science*, 2013; 341:387–91.

144. Wykes R. C., Heeroma J. H., Mantoan L., et al., 'Optogenetic and potassium channel gene therapy in a rodent model of focal neocortical epilepsy', *Science Translational Medicine*, 2012; 4:161ra52.

145. Wykes R. C., Kullmann D. M., Pavlov I., Magloire V., 'Optogenetic approaches to treat epilepsy', *Journal of Neuroscience Methods*, 2016; 260:215–20.

146. Fan Z. L., Wu B., Wu G. Y., et al., 'Optogenetic inhibition of ventral hippocampal neurons alleviates associative motor learning dysfunction in a rodent model of schizophrenia', *PLoS One*, 2019; 14:e0227200.

147. Barnett S. C., Perry B. A. L., Dalrymple-Alford J. C., Parr-Brownlie L. C., 'Optogenetic stimulation: understanding memory and treating deficits', *Hippocampus*, 2018; 28:457–70.

148. Fonseca R., Nagerl U. V., Morris R. G., Bonhoeffer T., 'Competing for memory: hippocampal LTP under regimes of reduced protein synthesis', *Neuron*, 2004; 44:1011–20.

149. Barron H. C., Vogels T. P., Behrens T. E., Ramaswami M., 'Inhibitory engrams in perception and memory', *Proceedings of the National Academy of Sciences of the USA*, 2017; 114:6666–74.

150. Edwards C. J., Suchard M. A., Lemey P., et al., 'Ancient hybridization and an Irish origin for the modern polar bear matriline', *Current Biology*, 2011; 21:1251–8.

151. Heinz T., Pala M., Gomez-Carballa A., Richards M. B., Salas A., 'Updating the African human mitochondrial DNA tree: relevance to forensic and population genetics', *Forensic Science International: Genetics*, 2017; 27:156–9.

152. O'Leary N., Jairaj C., Molloy E. J., McAuliffe F. M., Nixon E., O'Keane V., 'Antenatal depression and the impact on infant cognitive, language and motor development at six and twelve months postpartum', *Early Human Development*, 2019; 134:41–6.

153. McGovern D. P., Astle A. T., Clavin S. L., Newell F. N., 'Task-specific transfer of perceptual learning across sensory modalities', *Current Biology*, 2016; 26(1):R20–21.

154. Noh K., Lee H., Choi T. Y., et al., 'Negr1 controls adult hippocampal neurogenesis and affective behaviors', *Molecular Psychiatry*, 2019; 24:1189–205.

155. Frankland P. W, Bontempi B., 'The organization of recent and remote memories', *Nature Reviews Neuroscience*, 2005; 6:119–30.

156. Laureys S., Owen A. M., Schiff ND., 'Brain function in coma, vegetative state, and related disorders', *Lancet Neurology*, 2004; 3:537–46.

157. Kelly J. R., Baker A., Babiker M., Burke L., Brennan C., O'Keane V., 'The psychedelic renaissance: the next trip for psychiatry?', *Irish Journal of Psychological Medicine*, 2019:1–5.

References

158. O'Keane V., O'Hanlon M., Webb M., Dinan T., 'd-fenfluramine/prolactin response throughout the menstrual cycle: evidence for an oestrogen-induced alteration', *Clinical Endocrinology* (Oxford), 1991; 34:289–92.

159. O'Keane V., Dinan T. G., 'Sex steroid priming effects on growth hormone response to pyridostigmine throughout the menstrual cycle', *Journal of Clinical Endocrinology and Metabolism*, 1992; 75:11–14.

160. Frokjaer V. G., Pinborg A., Holst K. K., et al., 'Role of serotonin transporter changes in depressive responses to sex-steroid hormone manipulation: a positron emission tomography study', *Biological Psychiatry*, 2015; 78:534–43.

161. Dinan T. G., Cryan J. F., 'The microbiome–gut–brain axis in health and disease', *Gastroenterology Clinics of North America*, 2017; 46:77–89.

Index